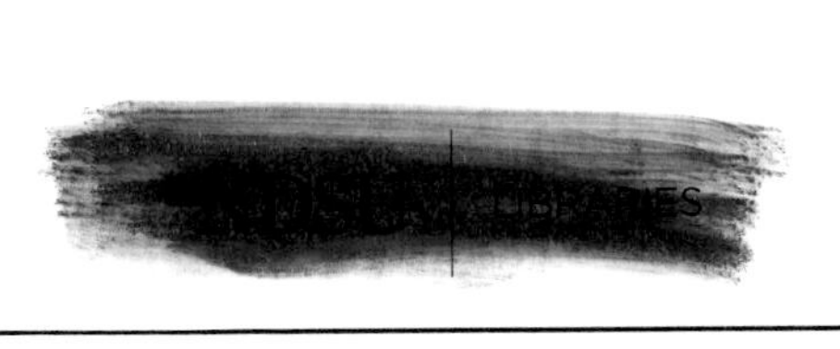

The Social Turn in Moral Psychology

The Social Turn in Moral Psychology

Mark Fedyk

The MIT Press
Cambridge, Massachusetts
London, England

This book was set in Stone Serif and Stone Sans by Toppan Best-set Premedia Limited. Printed and bound in the United States of America.

Library of Congress Cataloging-in-Publication Data

Names: Fedyk, Mark, author.
Title: The social turn in moral psychology / Mark Fedyk.
Description: Cambridge, MA : MIT Press, [2017] | Includes bibliographical references and index.
Identifiers: LCCN 2016022298 | ISBN 9780262035569 (hardcover : alk. paper)
Subjects: LCSH: Ethics. | Ethics--Psychological aspects. | Moral development. | Psychology and philosophy. | Social psychology.
Classification: LCC BJ45 .F43 2017 | DDC 170.1/9--dc23 LC record available at https://lccn.loc.gov/2016022298

10 9 8 7 6 5 4 3 2 1

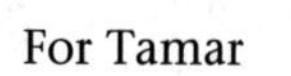
For Tamar

Contents

Acknowledgments ix

1 Introduction 1
2 Biology and Evolutionary Moral Psychology 23
3 The Social and the Psychological Dimensions of Ethics 49
4 Methodological Lessons from Philosophical Ethics 69
5 Inconsilience between ADJDM Moral Psychology and Philosophical Ethics 89
6 Moral Dumbfounding 117
7 The Causal Theory of Ethics 133
8 Hard-Question Social Theory 159
9 Descriptive Moral Psychology 181
10 Normative Moral Psychology 203
11 Conclusion 219

Index 233

Acknowledgments

It is odd that we credit individual people for producing academic research when the whole process is inherently collective. No academic research, great or small, ever comes into being but by various patterns of cooperation, collaboration, criticism, and resource sharing. That said, I of course am solely responsible for the errors, mistakes, and omissions in this book. But the following people deserve real credit for helping to make many of the good and worthwhile bits: Michel Shamy, Roopen Majithia, Jane Dryden, Robbie Moser, Khash Ghandi, Barbara Koslowski, Richmond Campbell, John Campbell, Jason Oakes, Eli Gerson, and the wonderful students at Mount Allison University in Sackville, New Brunswick, who have taken my classes over the years. Thank you all.

And then, special thanks are due to David Danks, Ian MacKay, and Nadia Chernyak, each of whom read and commented on several early drafts. I also want to thank Paul Teller for his support, and acknowledge how his ability to say even the most difficult philosophical ideas using only the right—and never any wrong or unnecessary—words has helped make this book better than it would otherwise be. Finally, I am so very deeply grateful to Richard Boyd for all of his support over the years. Readers familiar with his work will see clearly the influence of his thinking on my own; that, in my view, is a profoundly good thing.

Audiences at Mount Allison University, the University of Lisbon, Paris-Sorbonne University, Humboldt University of Berlin, Queen's University at Kingston, UC Berkeley, Vilnius University, Cornell University, VU University Amsterdam, and the University of Utah provided extremely useful questions, comments, and criticisms.

Jim Griesemer and Roberta Millstein helped out in the kind and humane way that will be familiar to anyone who knows them, by, among other things, opening up their Philosophy of Biology Lab at UC Davis to me for a semester. This provided me with the (always necessary when doing philosophy) social space to try out arguments for several of the more unusual ideas that can be found in the middle chapters of this book. I would like to thank, too, the student members of the Griesemer/Millstein lab. And I must also thank the first philosopher I met in person, Stephen Leighton. Many of the ideas in the following book are my attempt to answer just a few of the truly profound questions that he asked during his seminar on moral psychology in 2003.

My appreciation also to Phil Laughlin and the team at MIT Press for their assistance and guidance, and, finally, thanks to the three anonymous reviewers for MIT Press for their fantastic and incisive feedback.

1 Introduction

It has never been believed that moral psychology is methodologically self-sufficient. Progress in moral psychology is—and always has been—sustained by information first generated by different auxiliary disciplines and then imported into moral psychology. Philosophical ethics is among the most important of these disciplines, because moral psychology receives a metaphysical map of its own subject matter and many of its most useful analytical concepts from philosophical ethics. It is also widely believed that evolutionary biology is nearly as important to moral psychology as philosophical ethics, and perhaps even more so. No science to emerge in the last one hundred and fifty years has been as successful as evolutionary biology—and if certain tenets of evolutionary theory, or facts about the evolutionary history of our species, can be used to generate psychological hypotheses for testing in moral psychology, then the antecedent reliability of evolutionary biology can perhaps be transmitted to moral psychology.

How should we understand the relationship between moral psychology and the social sciences? A historical answer to this question is that, for at least some social sciences, moral psychology is prior to the social sciences. For example, soon before observing that the same view was held by J. S. Mill, J. N. Keynes writes: "political economy should then be described as a social, rather than as a moral or psychological, science. It presupposes psychology just as it presupposes the physical sciences ... but the science is not therefore a branch of psychology" (Keynes 1891, 85). The moral and the psychological are both prior to the social for the elder Keynes and Mill. Yet there is another answer to this question, an answer that is defended over the following chapters: the social sciences are prior to moral psychology; although putting the idea in those words is not exactly right. The alternative answer found in this book is that the social sciences and moral psychology can and should interact so that they inform one another in a mutually supporting fashion—but that, nevertheless, empirical research in moral psychology should begin with data that can only be provided by the social sciences, and in that specific sense the social sciences should be prior to moral psychology.

Put succinctly, then, this book argues that moral psychology should make a social turn. What does that look like? The basic idea is that moral psychology make its explanandum the psychological processes that motivate patterns of social behavior defined as *ethical* according to a normative theory that, in turn, relies partly on social and historical data, and partly on concepts from philosophical ethics, to determine which existing patterns of social behavior are indeed *ethical* patterns of behavior. Rather than defining ethical behavior as, for instance, just those actions that flow from moral judgments, where the construct *moral judgment* itself receives an a priori definition, the proposal is that an empirical theory be used to locate ethical behavior in different social worlds, and that moral psychology see as one of its core tasks the scientific investigation of the psychological states associated with producing the relevant social behaviors. More simply: social-turn moral psychology uses normative information extracted from the social sciences to determine which specific psychological hypotheses to investigate. Consequently, what makes a judgment a moral judgment can sometimes be determined by facts about the social context of the judgment, and not the propositional content or motivation force of the judgment itself.

That surely sounds fine in the abstract. However, to show more of what it looks like to undertake social-turn moral psychology, let me introduce a working example to help bring more of the theory's details and implications into view.

In 1969, Cornell University began an institutional—that is, "structural"—reorganization. A university senate was formed for the first time, and a new set of legal policies and procedures was designed for the campus, replacing the old and unpopular university judicial system. A controversial university president was replaced by a new leader who enjoyed more student support. There was, in so many words, a small revolution at Cornell. One of the events that precipitated this revolution was the occupation of a main campus building, Willard Straight Hall, by students from the Afro-American Society (AAS) on the night of April 19 earlier in the same year. Soon after seizing Willard Straight, the occupying students were confronted by members of the Delta Upsilon fraternity, and, as a result of this confrontation, some of the occupying students armed themselves with guns. A UPI photographer won the Pulitzer Prize for his now-famous photo of the AAS students returning to Willard Straight with their firearms. And, at about the same time as the AAS students armed themselves, members of Students for a Democratic Society (SDS) formed a ring around Willard Straight Hall, a sizable building, lending their support to the members of AAS inside. Toward the end of the occupation, as police were preparing to force the students from the building, university officials and AAS leaders were able to negotiate a peaceful end to the standoff. Willard Straight Hall was occupied for about thirty-six hours.

Now, there are different descriptive levels at which we can explain both the takeover of Willard Straight and the political and social effects of the occupation. There are, first of all, structural explanations that refer, inter alia, to the institutional causes of the event. For instance, we can talk about the old legal system used at Cornell prior to 1969, and examine the different impacts its application had on either the general student body or different subpopulations. We can then compare the "population-level" or "community-level" impacts of the old legal system with those produced by the new legal system. Of note is the fact that the members of AAS took the structural perspective when they planned their occupation of Willard Straight: their argument was that the legal system was in various ways unfair because of its intolerance of certain forms of political speech. Their objections did not refer to the psychological states of any particular individual at Cornell; and the students were certainly not asking for an apology from the university president or merely to be heard by the board of trustees. Put metaphysically, the AAS students were protesting the incompatibility of certain social behaviors and the existing framework of legal norms at Cornell.

We can also describe the causes and effects of the takeover at the level of individual psychology. We can ask what beliefs and desires motivated the protesters, their allies, and their enemies on the night of the takeover. We can also look at what beliefs were formed, or changed, in observers of the takeover in the broader university community—and perhaps also how these beliefs influenced key decision makers to alter the institutional structure of Cornell. The important point to understand, however, is that the two levels of description—structural and psychological—are not mutually exclusive. Any competent history of the takeover will integrate both levels of explanations; see Downs 1999 for an example of just this.

What is the relevance of the Willard Straight takeover to moral psychology? There is an important methodological lesson that we can explore here. It is tempting to see the debate surrounding the occupation of Willard Straight as an example of political and ethical conflict, and in so doing use the languages of interest groups, race relations, and even Cold War ideology to describe what occurred in the spring of 1969 at Cornell. Using any of these conceptual vocabularies would portray the conflict as enmeshed with an even larger and less tractable debate. And it very much was that, but it was also something else, too. What is of much more importance in light of the arguments to follow is that the occupation, and the debate surrounding it, belies the presence of a fundamental commitment to a basic and extremely common form of social cooperation among the disputants—specifically: the shared willingness to maintain a particular normatively integrated community. It is true that at dispute were the fundamental norms regulating undergraduate student life at Cornell, both the de jure principles used to determine official sanctions and rights for students; and the broader set of de facto norms that played a role in determining how different social groups at Cornell were treated, sanctioned, or differentially favored by the administration. But beneath this dispute was a shared willingness to maintain the existence of a population defined as such by its norms. And then, the less obvious metaphysical point to see here is that such a population can continue to exist *even when* its most fundamental norms are in dispute. In our example, the relevant normative population had no official name, so it can only be called somewhat loosely the "Cornell community." But again, the key point here is to see that, for all of the disagreements in this community, there was no disagreement over whether the Cornell community should continue to exist—and that particular fact explains in part why the dispute was so contentious. The outcomes of the dispute had genuine importance to different groups in the community because it was simply not possible to opt out of an unfavorable resolution by either leaving or dissolving the community. Generally, then, normative debates can be as difficult and intractable as they often are precisely because these debates frequently occur where there is a shared commitment to preserve an existing normatively integrated population. So, the presence of normative disputes should not be mistaken for the absence of a normatively integrated population, and moral psychologists should not let their focus rest entirely on either seemingly or genuinely intractable moral problems.

We have already noted that the takeover of Willard Straight shows that the history of at least normatively integrated populations can be understood using both structural and psychological explanations. More importantly, however, we must also note that reference to phenomena at the structural level of analysis is necessary in order to draw conclusions about the effects of adopting or rejecting any number of different norms on the outcomes of different groups in the Cornell community, and whether or not these effects improved or worsened or made no difference for the community. It is not a psychological question to ask whether or not the changes at Cornell increased procedural fairness; more generally, determining whether some population is doing better or worse over the course of its history can require investigating any number of structural phenomena, where concepts useful for referring to these phenomena cannot be translated into, or reduced to, a set of corresponding psychological concepts.

It also follows from this lesson that the conceptual vocabulary of moral psychology must not be asked to do more work than it can do; telling the moral and ethical history of different normatively integrated populations is *not* among the tasks of moral psychology, if the inductive and explanatory usefulness of its proprietary concepts is to be preserved. There is work to be done, then, exploring how best to integrate the different levels and kinds of normative explanation and analysis, whether or not our goal is only to account for patterns of past social behavior, or it is also to provide guidance about future behavior.

The argument to be developed hereafter takes for granted that the different levels of analysis—structural and psychological—that we have applied to the revolution at Cornell in 1969 are hardly unique. Since it is possible to analyze the effects of institutions at the structural level, I will argue that research in this area provides a natural starting point for new research projects in moral psychology. Indeed, and as we shall see, it is routinely the case that structural explanations in the social sciences reveal how social institutions—and other structural components of different human populations—cause or otherwise affect outcomes for groups of people that are, in different ways, evaluatively valenced. This observation leads to the proposal that moral psychology—or, rather, one kind of moral psychology (as discussed in chapter 9)—can be about the psychological basis of those social behaviors that evidence from the social sciences helps to show are *ethical* behaviors, because the social scientific evidence demonstrates that the behaviors are an important cause of a valuable outcome.

Of course, in order to make that inference, we need to rely on a theory of ethics that has an important property: the ethical theory must pick out, and do so on principled grounds, patterns of actual ethical behavior that exist at the structural level in real human populations. A theory with just such a property is introduced in the middle chapters of this book—and the intuition it rests upon is that a social behavior is ethical if (and *not* only if; there are many other reasons why some social behavior can be ethical) there is evidence that it is necessary to produce some specific valuable structural outcome. Crucially, this ethical theory is useful to moral psychology in a way that even some of the best theories in philosophical ethics are not. The underlying problem is that most ethical theories in the history of moral philosophy are, technically, *unrealized*. I will explain more about what I mean by "unrealized" presently; the idea in its simplest form is that for an ethical theory to be unrealized is for there to be no normatively integrated populations that has put the ethical theory into practice. And the fact that even the best theories in philosophical ethics are unrealized is a serious methodological obstacle for moral psychology, because it means that the discipline cannot rely on rational estimates of the plausibility of theories in philosophical ethics to determine which populations to sample from, or attempt to generalize to, in order to assess the external validity of any of its empirical research. At its worst, then, the problem is that moral psychology must either choose between philosophical plausibility or empirical tractability—and the ethical theory presented in the chapters 7 and 8 of this book is designed to help moral psychology avoid the horns of that dilemma.

Let us return one last time to the revolution at Cornell to make these various ideas a little more clear. I will argue that what makes the changes brought about by the Willard Straight takeover a sequence of *ethical* changes is not just that they were seen as such by members of the community at the time. What makes these specific changes ethical is that they were *structural* changes that occurred within a normatively integrated population, and, crucially, that these changes had several different but concrete good and valuable social outcomes for different groups in the Cornell community. The fundamental idea behind social-turn moral psychology, thus, is that evidence that certain structural elements of normatively integrated populations cause different evaluatively valenced social outcomes provides a starting point for inquiry in moral psychology. I will argue, specifically, that one (and not the only) core task for moral psychology is investigating the psychological dimensions underlying the relevant structural causes and beneficial effects. So, for example, a program of research in social-turn moral psychology might try to understand how learning about the Willard Straight takeover changed people's attitudes about campus legal codes, or whether the takeover led to increasingly polarized attitudes about racial politics, or if most people who supported the legal changes would be able to explain the rationale behind the changes without referring to the vivid events of the takeover, or whether people who opposed the takeover were more likely to support Richard Nixon and Spiro Agnew, because the latter frequently referred to the takeover in his speeches. Or, to put the same idea generically: social-turn moral psychology makes its focus the psychological foundations of macrolevel social behaviors that are known on independent grounds to be the cause of good (or bad; though I will not typically focus on this possibility) outcomes for different normatively integrated populations.

So, what's another core task for moral psychology? It is not my view that the *only* things of value are the different kinds of social outcomes that can be explained by referring to the normative structure of the population experiencing the relevant outcomes. These are only one type of good and valuable outcome—and a completely separate category captures the properties of individual people (or, more precisely, properties of individual lives) that are valuable. I will argue (see chapter 10) that a separate task for moral psychology is the study of the causes of these personal, nonstructual goods—goods like a experiencing a meaningful life, increasing sense of one's agency, and so on. But this line of inquiry can only be seen for what it is once the various details of social-turn moral psychology are laid out in detail.

Why Should You Care?

Why should philosophers and psychologists invest the effort necessary to understand the following arguments? The conceptual possibility of making the social turn in moral psychology is not a reason to learn about the details of the turn, let alone invest the real time, energy, and other resources that implementing the social turn, even just in approximation, would involve.

To address this concern, the first part of this book develops the thesis that there are problems in methods that are increasingly central to moral psychology and that, if left unchecked, may threaten the reliability of the discipline. As mentioned earlier, there is a consensus view that moral psychology depends on both philosophical ethics and evolutionary biology. However, the first third of the book (chapters 2–6) demonstrates that in a number of important ways actual practice in moral psychology is not consistent with the consensus view. The specific ways that moral psychology relies on evolutionary biology are, in fact, incompatible with certain of the most general methodological norms of evolutionary biology (see chapter 2), and because of this there is evidence of growing inconsilience between moral psychology and evolutionary biology. Likewise, there is evidence of inconsilience between philosophical ethics and moral psychology (chapters 4 and 5) that stems from attempts in moral psychology to borrow concepts that have been historically central to moral philosophy, and in the process, redefine these concepts so that they apply to that kinds of psychological phenomena that are measurable in the context of a psychological experiment. Thus, the argument to come is not that the consensus view should be rejected and replaced with social-turn moral psychology. Rather, the argument is that social-turn moral psychology is a better way of realizing the consensus view, because implementing the social turn in moral psychology is a way of improving consilience among moral psychology, philosophical ethics, and evolutionary biology. Readers should therefore care about the following arguments and analyses if they want a framework that represents a better way of integrating these three disciplines.

Here, then, in more detail is how the following chapters are organized. Chapter 2 shows that it is not possible to derive psychological conclusions from evolutionary explanations, and in so doing introduces a principle, called "Mayr's lemma," which will be used in many of the remaining chapters. This principle entails that it is not typically possible to derive structural explanations from psychological explanations, and vice versa. However, since many contemporary moral psychologists think that the way moral psychology is related to evolutionary biology is that different kinds of structural explanations (more accurately: ultimate explanations that refer to a process of natural selection) in evolutionary biology can be used to make predictions about the existence of specific but otherwise unknown psychological mechanisms, chapter 2 opens up a new question: how then should evolutionary biology be related to moral psychology? The answer I defend is that moral psychology should strive for methodological consilience with evolutionary biology by using methods that do not violate Mayr's lemma. That means, practically speaking, keeping structural and psychological explanations of any given pattern of normative behavior separate and distinct from one another.

What about consilience between philosophical ethics and moral psychology? Chapter 3 proposes an analytical framework that provides a way of explicitly analyzing the relationship between philosophical ethics and moral psychology. A very simple way of stating the idea underlying the proposal in chapter 3 is that moral philosophers have often been in the business of evaluating rules that define the structure of possible ideal social games, without concern for whether or not the games ever are, or ever can be, played. Moral psychology could therefore be said to be in the business of figuring out how people do, or could, or even whether they ever can, learn to play these games. What emerges from the analysis in chapter 3, crucially, is a small cluster of concepts that allow us to state clearly and simply what the relationship between philosophical ethics and moral psychology can be.

Chapter 4 then answers the question: what are some methodological principles that moral psychology should follow if it is to remain consilient with philosophical ethics? Three such principles are described in chapter 4, and then chapter 5 demonstrates that these principles are routinely violated by some of the most influential contemporary research in moral psychology. Chapter 6 then deals with an objection to the argument advanced through chapters 2 and 5. By the end of chapter 6, we will have a full view of the problem that motivates this book: moral psychology is becoming increasingly inconsilient with both evolutionary biology and philosophical ethics.

The remaining chapters chart a way out of this problem. The underlying strategy is simple: construct an ethical theory that is philosophically plausible and useful to moral psychology. In other words, the strategy is to use concepts from philosophical ethics to describe different patterns of existing social behavior that, if realized, are ethical—and then to show how moral psychologists can study the psychological foundations and dynamics of these social behaviors. To return to a point mentioned briefly earlier, one of the reasons why moral psychology and philosophical ethics are becoming inconsilient is that the most plausible theories in philosophical ethics are unrealized, and consequently these theories provide moral psychology with no meaningful guidance about which populations to either sample from or generalize to for the purpose of validating research in moral psychology. Of course, making the observation that philosophical ethics is mostly not about existing patterns of social behavior is not to criticize philosophical ethics—the point is only that this observation can explain why in recent years many moral psychologists have frequently redefined concepts borrowed from philosophical ethics so that they can apply to existing patterns of thought and behavior. But the general thrust of the argument in chapters 4 and 5 is that these redefinitions have been much too severe, such that the revised concepts have lost the projectable—in the sense given to that term by Nelson Goodman (1955)—meanings they originally had in virtue of their entrenchment in the vocabulary of philosophical ethics.

Chapters 7 and 8 develop a theory of ethics for moral psychology that does not create tension between its scientific practice and the conceptual vocabulary of philosophical ethics. This theory—called the "causal theory of ethics"—provides a principled way of determining which social behaviors are ethical behaviors by telling us which kinds of social theory (chapter 8) count as descriptions of ethical behavior. The causal theory also provides a conceptual framework that is projectable from philosophical ethics. The fundamental suggestion, then, is that the causal theory of ethics can be used to identify in the social sciences what eventually becomes called "hard-question social research," and that the microfoundations of the relevant patterns of social behavior described by any such hard-question research are the proper object of investigation for descriptive moral psychology.

Two Guiding Philosophical Assumptions

Every philosophical and scientific project must begin by adopting a cluster of philosophical assumptions that can only be adequately defended by another independent line of inquiry. And there are two fundamental philosophical assumptions that support this book: moral naturalism and conceptual framework pluralism. It will help to briefly explain how I understand both of these ideas here in the introduction to this book.

Let's deal with moral naturalism first. Most of the following arguments take for granted that morality has distinctive social functions. These functions provide the grounds for evaluating the rationality of morality on instrumental grounds. Indeed, we will consider evidence in the chapters 7, 8, and 10 that indirectly provide support for the following thesis: it is rational for morality to continue to exist because, even if only imperfectly, sometimes it meets some of its most lofty goals. Here is a comparison to help make the idea clear. Science has any number of internally rational goals; paramount among these is probably the goal of producing increasingly refined causal knowledge. At the same time, there simply is no argument that can persuade absolutely anyone, no matter what their prior beliefs or experiences happen to be, that science is a worthwhile activity, or that any such a person should strive to be a scientist. What justifies scientific practice to its practitioners is nothing more and nothing less than the ability over decades for scientists to make progress toward achieving some of their goals. To see that point, imagine that scientific practice only ever leads to profound mistakes about the operation of nature's causal systems, and otherwise fails utterly to make progress toward any of its other ends. Sooner or later, scientists would simply give up their practice—and while there would still be a historical (or even evolutionary) story to tell about how scientific practice once came to be widespread, there would be no further argument that would persuade people to take up the practice again. The moral naturalism that I am taking on board is of a similar character: morality has several internally rational ends, and there is evidence, in both history and contemporary social science, that morality has made progress toward realizing some of these ends. This progress can justify morality for those who already understand the point of its practice. So, the same pattern of argument that justifies scientific practice also justifies the existence of morality—and does so without reference to necessary truths, deities, or the thesis that moral knowledge is inherently a priori or is delivered entirely by some kind of special moral intuition.

But what are the internally rational ends of morality? Later chapters contain several concrete examples, and what these examples all have in common is that they describe certain social behaviors that, because of their good effects, should be preserved, or otherwise have priority over other practices if and when a conflict arises. Relatedly, I will argue that it is rational for normative communities to make certain patterns of social behavior obligatory, or, what amounts to the same thing, to give certain norms trumping power over all other norms when it comes to motivating patterns of social behavior. The quick justification for this thesis is as follows. Humans are, fundamentally, normative. Even very young children have the ability to reason about norms, perceive and reason about normative violations, and perhaps even recognize and think through situations where two or more norms conflict with one another (Schmidt, Rakoczy, and Tomasello 2011, 2012; Schmidt and Sommerville 2011; Wyman, Rakoczy, and Tomasello 2009). And in any practical context, there are infinitely many norms that a person could follow, and, partly because of this, a particular kind of normative uncertainty is a ubiquitous feature of social life. We are always forced to make choices about which norms to follow and not to follow when we act, and we rarely have conclusive evidence that anyone who is not a family member, friend, or colleague will make the same choices that we will make about which norms to follow. When these conflicts occur, and when one particular norm is followed and some others are not, we can say that the first norm *trumps* the other norm. So, both Wilson and Sarah may think that norms A, B, and C should be followed in certain social contexts, but it is also possible that Wilson thinks that A should trump B and C, while Sarah thinks that C should trump A and B. Given these different orderings, Wilson and Sarah will fail to have coordinated behavior in a situation where it is possible to follow A, B, and C. However, note that the practice of making certain norms as obligatory—that is, by requiring that certain norms trump *all* other norms—provides a solution to the problem. If C is obligatory, then no matter how Wilson and Sarah rank A and B relative to each other, both will know what the other will do. The proposal, then, is that one of the functions of morality is that it helps solve certain basic social coordination problems—and I will propose that what makes a norm an *ethical* norm is that it either does or should trump all other norms that could apply to some context of action. That is not the whole story, of course. It is even better if the norms that are defined as ethical create when realized social conditions that make possible other more intricate patterns of cooperation and sociality; and it is even better if we can figure out what further norms would, if realized, produce various different good and valuable social outcomes, and when possible fix those norms as ethical too in order to lock in their good effects. But never mind these more speculative ideas for now, as the underlying thesis should be clear: the moral naturalism that I accept holds that morality is rational because morality establishes a cooperative platform upon which other forms of social interaction can develop, and many of these developments are valuable when seen from even a minimally moral perspective. And the coordinating function of obligatory norms is one of the most basic functions of morality that I suggest provides people who buy into morality with a reason to stick with the practice.

More generally, it should be becoming clear that the moral naturalism that informs this book is one that seeks to hold the middle ground between, on one side, skepticism about the rationality of morality often associated with scientific theories of morality and, on the other side, the conviction that moral knowledge springs from some uniquely reliable epistemic source, and that moral knowledge is therefore in some way is categorically different from all other kinds of knowledge of the natural world. And of course, middle-ground views like my moral naturalism have been defended in the past by both philosophers and psychologists (Turiel 1990; Mackie 1990; Turiel, Killen, and Helwig 1987).

We turn now to an idea that I express as "conceptual framework pluralism": the idea behind the phrase approaches philosophical common sense in the philosophy of biology. Writing more than forty years ago, William Wimsatt observed that "in biology and the social sciences, there is an obvious plurality of large, small, and middle-range theories and models, which overlap in unclear ways and which usually partially supplement and partially contradict one another in explaining the interaction of phenomena at a number of levels of description and organization" (Wimsatt 1972, 67). The explanation for why it is easy to find evidence to support Wimsatt's generalization is that the complexity of the systems renders them unknowable if scientists are also required to only use a single (inferentially, or mathematically, or logically closed) conceptual vocabulary in order to build theories of the relevant system. Instead, to manage the complexity, scientists deploy families of models that are able capture information about different aspects or dimensions of these complex systems. Crucially, models are never perfect representations of the aspects of the systems they represent (Teller 2001; Giere 2004). So, because models are only ever rough approximations, it is possible for several different models that are mutually inconsistent to nevertheless carry information about different dimensions or aspects of one otherwise unified natural system. Practically speaking, this requires that scientific expertise include knowledge of how to apply and move between these models; James Griesemer and his coauthors sometimes call this knowledge a scientist's "theoretical perspective" (Caporael,

Griesemer, and Wimsatt 2013). This theoretical perspective is a (sometimes only implicit) body of metaphysical knowledge that helps a scientist use her models to generate explanations or predictions that satisfy, inter alia, the epistemic goals of her area of inquiry, and it too is only ever an approximation of the system that is being modeled. It can be thought of, if only intuitively, as the implicit model that tells a scientist how to use her explicit models. However, the more important point is that one strategy that scientists have successfully employed for generating knowledge of complex systems is to develop a plurality of different conceptual frameworks that are expressed in otherwise incompatible and sometimes even literally inconsistent models—and, in parallel with this, scientists have also developed the different forms of expertise ("theoretical perspectives") that are required in order to know how to use the different models. And what I will refer to with the phrase "conceptual framework pluralism" is the successful development of a plurality of different conceptual vocabularies useful for generating useful models of some particularly complex natural system.

Here is a test that can be used to determine whether or not someone accepts conceptual framework pluralism. Suppose we have three different mid-level models X, Y, and Z of some system. Suppose that these theories are mutually inconsistent, but at the same time, each theory has been used for over twenty years to address a range of different empirical problems. A scientist who rejects conceptual framework pluralism will ask, for instance, which of X, Y, and X is "really" true, and may spend her time seeking evidence that can falsify two of the theories. By contrast, a scientist who accepts conceptual framework pluralism will instead look for insights and practices that reflect intelligent ways of applying X, Y, and Z to solve increasingly deeper or more advanced problems. She will not care which of these theories is "really true"—instead, she will try to develop an implicit metaphysical theory that helps her understand what phenomena X, Y, and Z are respectively mostly likely true of. She understands that literal interpretations of, say, the differential equations in Y are naive; but that X, Y, and Z are still pretty good representations of the system she cares to know more about. Each theory is, in its own way, a serviceable approximation of an aspect of a very complex whole.

Now, one of the fundamental arguments in this book is that conceptual framework pluralism can be applied to the study of the normative behavior of humans. My fundamental aim is to show that there is a way of combining various different conceptual vocabularies—some from philosophical ethics, others from different branches of social and moral psychology—into a larger metaphysical theory of human normative behavior. I stress, moreover, that this metaphysical picture is not reductive or eliminativist—the possibility of real ethical progress, and deep normative error, is preserved in the picture. And so, for example, I am not interested in the question of whether some formulation of virtue ethics is true, and all formulations of deontology therefore false. Instead, I want to see if it is possible to understand enough about the empirical metaphysics of normative behavior in humans so as to be able to isolate, for instance, the specific aspects of normative behavior that the conceptual vocabulary of virtue ethics best applies to, and then what other aspects (if any) of normative behavior deontological ethical theories best apply to, and finally, what implications for research in moral psychology emerge from this picture.

The path lit by conceptual framework pluralism is not well trod in moral philosophy. Indeed, among contemporary moral philosophers, there instead seems to be a nearly universal underlying assumption that the right or correct normative theory must be expressible in a single, unified—and perhaps even deductively closed—conceptual vocabulary. Remarks by Kate Manne get the relevant idea across well enough: "Bodily imperatives are a promising candidate for constituting the core claims of morality—in this case, to make pain stop, or better yet, to prevent it from occurring in the first place. And, once this move has been made, we can say that desires to violate the bodily imperatives which constitute core moral claims are morally objectionable" (Manne 2015). Later on, she continues: "For those of us who are secularists, and believe that the world is all we have, we must find morality within it. We must stop waiting for Godot. This is a lesson taught to us over half a century ago by Anscombe, in her 'Modern Moral Philosophy.' But it still bears repeating, I think" (Manne 2015).

Manne's invocation of Anscombe indicates that here the question is where in the world a naturalist should locate morality. Bodily pains are of course within the ontology of any moral naturalist, and Manne describes in very fine analytical detail how building out from them can lead us to traditional moral norms. But the alternative presented here—a theory of ethics and morality informed by conceptual framework pluralism—is to reject the idea that there are "core claims" of morality, because we should not assume a priori that a single conceptual vocabulary and its associated inferential framework are sufficient to express all of the normative truths we may otherwise find evidence for. Perhaps our total moral and ethical knowledge may be expressible only if we first assemble a loose confederation of different conceptual vocabularies, some of which apply to large-scale social phenomena, and others to deeply personal phenomena. Of course, it may be that individual philosophers are able to develop the theoretical perspective necessary to hold these different conceptual frameworks together—but we should also take seriously the possibility that normative behavior in humans is so complex that, as is the case in many scientific fields, the expertise required to organize the different vocabularies into a useful and reasonably general picture requires the formation of a cooperative community of experts with their own distinct set of epistemic norms.

That said, a single book cannot hope to account for, let alone evaluate, the multitude of different conceptual vocabularies that may eventually prove to be apt for describing aspects of the normative behavior of our species. Please do not misread me as also arguing that this is the only way, or even among the best ways, of constructing a pluralist framework suitable for understanding, critiquing, and maybe even directing normative behavior. I would be very happy if philosophical criticisms of this book accept both conceptual framework pluralism and moral naturalism, and demonstrate the errors of my work by charting out a superior metaphysical theory of how to integrate philosophical and scientific research into an organized understanding of human normative behavior.

Realized Norms and Structural Explanations

A key analytical concept that is used in almost all of the following arguments is the concept of a *realizable norm*. This concept is significantly different from how norms are usually conceptualized in normative philosophy; as I understand it, the concept is derived from work in ethnomethodology, and perhaps the clearest way of illustrating the concept is to briefly return to Harold Garfinkel's famous breaching experiments (Garfinkel 1991). The idea behind these experiments is that a good way of understanding a social institution—which are complexes of realized norms—is to try to "break" the social institution by deliberating engaging in normatively inappropriate behavior. Responses to normatively deviant actions and choices can then be used to understand the content and contours of the different layers of realized norms that structure and regulate social life. Some of Garfinkel's students were sent to their parental homes, where they acted as if they were borders, not family. Other students collected their observations by engaging in conversations with friends as if they had deep, hidden motivations. A game of tic-tac-toe was invented in which after the first player had made their move, the other player erased the first player's mark, drew it into another square, and then proceeded to make a mark of their own. In these and most of the other breaching experiments, the typical response was a mixture of outrage, frustration, anger, and bewilderment. Yet these reactions to the deviant behaviors provide compelling evidence about the content of the realized norms of family life, tic-tac-toe, and conversation among friends. Simply put, then, realized norms are the norms that create and sustain social order. Consequently, these are the norms for which there is a strong prima facie expectation that they will be followed by members of distinct normatively integrated communities.

Here, then, is an analytical framework for understanding the concept of a realized norm. Any norm is either realizable or unrealizable; realizable norms may or may not be realized; unrealizable norms cannot possibly be realized; and when a norm is realized (whether individually or in conjunction with other norms) then a statement of the norm is also a true description of a pattern of social behavior. Intuitively, then, the norms of most sports are realized, and therefore realizable, norms. The norms of a newly invented sport that no one has yet thought to play are unrealized realizable norms. And the categorical imperative is an unrealizable norm, because strictly speaking, it cannot be a true description of any patterns of social behavior. But that is not to deny, of course, that the categorical imperative may function as a regulative ideal. In such a case, the norms that are realized will have different propositional content than that of the categorical imperative.

Indeed, many of the norms that have been of traditional interest to moral philosophers are regulative ideals: it is taken for granted that the norms do not describe existing social behavior of any group or community. Nevertheless, the expectation is that regulative ideal norms are a goal toward which a community may progress, such that its members' behavior becomes an increasingly better approximation of what it would be to realize the relevant norms. Yet, there are two important reasons why a regulative ideal norm may be discovered to be unrealizable. The first is that, quite simply, humans may not be constituted so as to be able to put the norm into practice. To use a silly example, if the norm "always use your wings to fly to work" were a regulative ideal norm, then it would be unrealizable because humans cannot fly. The other far more important—and I believe underappreciated by contemporary moral philosophy—reason why some regulative ideal norm may be unrealizable is that social and normative change is path-dependent. There may be no possible set of changes from the realized norms of some community to a new normative configuration that counts as a pretty good implementation of a set of regulative ideal norms. The gravitational influence, as it were, of the proposed ideal norms are too weak to shift the mass and momentum of existing normative history and practice from whatever its trajectory may happen to be.

Let me return to a point I made earlier. In order to be able to assess the external validity of any of her theories, a moral psychologist needs a theory of which social behaviors are ethical. Because of its focus on unrealized and sometimes unrealizable norms, mainstream philosophical ethics cannot provide such a theory. That is why a new ethical theory is introduced in chapter 7; but here I want to forestall an objection. This theory should not be read as competing with more traditional theories in philosophical ethics. Keep in mind the implications of conceptual framework pluralism: the goal of the causal theory of ethics is to provide a theory that focuses on the transition between unrealized but realizable norms and realized realizable norms, and so focuses on the kinds of normative change that are compatible with the path-dependency of cultural history. As such, the theory should be read as complementing the many philosophical theories that focus on regulative ideal norms. Or, more plainly, the goal of the causal theory of ethics is certainly not to change the subject of philosophical inquiry, but instead, to expand (and not by very much) the range of phenomena that philosophical theories of ethics can be applied to. Consequently, most of the uses of "norm" over the following chapters should be read as referring to *realizable* norms only.

Finally, I have used the phrase "structural explanations" frequently throughout this chapter, and it may be helpful to say a few words about how I will be using this concept. Structural explanations are frequently contrasted with individualistic explanations. Suppose we want to explain some pattern of social behavior SB. A structural explanation of SB will refer to supra-individual properties of the population in which SB emerges—properties such as the laws of the country, how an individual can be positioned in a society in virtue of their (assigned, perceived, or chosen) social role, or how the distribution of a several properties in the population produces a number of distinct effects in the population as a whole, and so on. An individualistic explanation of SB will refer to the beliefs, desires, and other motivations components of the people who perform SB. So, whether economic growth in America between 1950 and 1960 was more equitable than growth between 1960 and 2000 is a structural question—even though, metaphysically speaking, steadily increasing the productive capacity of a national economy is mostly an (immensely complex) social behavior. But whether people usually value equitable growth is a psychological question, and so too is whether steady and equitable growth is preferred over rapid and volatile growth. However, whether steady and equitable growth can be realized in contemporary America is an immensely complex structural question, and that remains true even if to answer that question we must also find answers to any number of psychological questions about, for instance, the range of political and economic beliefs held by both elected representatives and nonelected officials.

From the perspective of conceptual framework pluralism, structural and individualistic explanations can complement and even inform one another. However, that is not the most common view about the relationship between these two categories of explanation. Many readers will be familiar with the debate over methodological individualism, and this kind of individualism is, in brief, the doctrine that all structural explanations should be replaced by psychological explanations when accounting the social behavior of humans (Jackson and Pettit 1992). A commitment to individualism remains quite common among economists, especially among scholars who accept the basic thrust of the Lucas critique (Lucas 1976). The underlying attitude has been nicely expressed by Kenneth Arrow, who writes "it is a touchstone of accepted economics that all explanations must run in terms of the actions and reactions of individuals" (Arrow 1994). But the doctrine of course predates Lucas; we find Ludwig von Mises writing that "the forces determining the—continually changing—state of the market are the value judgments of individuals and their actions as directed by these value judgments" (von Mises 1949, 257–258).

So, the last point to be made in this chapter is that conceptual framework pluralism and methodological individualism are in tension. A pluralist can happily accept what an individualist cannot—namely, that individualistic explanations may be extremely apt for some aspects of a particular social behavior, and yet certain other aspects of the very same behavior may only be well explained by structural accounts. But I will argue that individualism in moral psychology leads to inconsilience with philosophical ethics, and even more importantly, it turns moral psychology's focus away from a host of fascinating, and morally important, questions that only come into view when information gleaned from different social sciences is taken into account.

References

Arrow, Kenneth J. 1994. "Methodological Individualism and Social Knowledge." *American Economic Review* 84 (2): 1–9.

Caporael, Linnda R., James R. Griesemer, and William C. Wimsatt. 2013. *Developing Scaffolds in Evolution, Culture, and Cognition*. Cambridge, MA: MIT Press.

Downs, Donald Alexander. 1999. *Cornell '69: Liberalism and the Crisis of the American University*. Ithaca, NY: Cornell University Press.

Garfinkel, Harold. 1991. *Studies in Ethnomethodology*. 1st ed. Malden, MA: Blackwell Publishing.

Giere, Ronald N. 2004. "How Models Are Used to Represent Reality." *Philosophy of Science* 71 (5): 742–752.

Goodman, Nelson. 1955. *Fact, Fiction, and Forecast*. Cambridge, MA: Harvard University Press.

Jackson, Frank, and Philip Pettit. 1992. "Structural Explanations in Social Theory." In *Reduction, Explanation, and Realism*, ed. David Charles and Kathleen Lennon, 97–132. Oxford: Oxford University Press.

Keynes, J. N. 1891. *The Scope and Method of Political Economy*. New York: Macmillan and Co.

Lucas, Robert. 1976. "Econometric Policy Evaluation: A Critique." In *The Phillips Curve and Labor Markets*, ed. K. Brunner and A. Meltzer, 19–46. New York: American Elsevier.

Mackie, J. L. 1990. *Ethics: Inventing Right and Wrong*. New edition. New York: Penguin.

Manne, Kate. 2015. "Locating Morality: Moral Imperatives as Bodily Imperatives." *UNC Meta-Ethics Workshop*. http://metaethicsworkshop.web.unc.edu/files/2015/04/Mane-Locating-Morality.pdf.

Schmidt, Marco F. H., Hannes Rakoczy, and Michael Tomasello. 2011. "Young Children Attribute Normativity to Novel Actions without Pedagogy or Normative Language." *Developmental Science* 14 (3): 530–539.

Schmidt, Marco F. H., Hannes Rakoczy, and Michael Tomasello. 2012. "Young Children Enforce Social Norms Selectively Depending on the Violator's Group Affiliation." *Cognition* 124 (3): 325–333.

Schmidt, Marco F. H., and Jessica A. Sommerville. 2011. "Fairness Expectations and Altruistic Sharing in 15-Month-Old Human Infants." *PLoS One* 6 (10): e23223.

Teller, Paul. 2001. "Twilight of the Perfect Model Model." *Erkenntnis: An International Journal of Analytic Philosophy* 55 (3): 393–415.

Turiel, Elliot. 1990. "Moral Judgment, Action, and Development." *New Directions for Child and Adolescent Development* 47: 31–50.

Turiel, Elliot, Melanie Killen, and Charles C. Helwig. 1987. "Morality: Its Structure, Functions, and Vagaries." In *The Emergence of Morality in Young Children,* 155–243. Chicago: University of Chicago Press.

von Mises, Ludwig. 1949. *Human Action*. Auburn, AL: Ludwig von Mises Institute.

Wimsatt, William C. 1972. "Complexity and Organization." In *Proceedings of the Biennial Meeting of the Philosophy of Science Association,* vol. 1972, 67–86. Chicago: University of Chicago Press.

Wyman, Emily, Hannes Rakoczy, and Michael Tomasello. 2009. "Normativity and Context in Young Children's Pretend Play." *Cognitive Development* 24 (2): 146–155.

2 Biology and Evolutionary Moral Psychology

In biology patterns of behavior in different animals are frequently explained when some of the structural causes of the relevant patterns can be described in sufficient detail. The most common examples of this methodological truism are of course explanations that refer to natural selection. The conclusion that a trait T is an adaptation is a conclusion licensed almost entirely by knowledge of the structural interactions that occur between organisms with the relevant phenotype, the environment in which they live, and the frequency of other traits among (usually) conspecifics—and where the effect of these interactions is to increase the fitness of bearers of T. David Sloan Wilson puts the idea this way: "One of the most basic facts about evolution is that fitness is a relative concept. It does not matter how well an organism survives and reproduces, only that it does so better than other organisms bearing alternative traits." (D. S. Wilson 2004, 245). So, facts about the structure of a population—specifically, facts about the distribution and frequency of a set of alternative traits within the population—must be known in order to construct an explanation in terms of natural selection. Furthermore, natural selection explanations are not the only form of structural explanation in biology: migration patterns, genetic drift, and many other kinds of evolutionary changes are commonly the subject of structural explanations in biology as well. In evolutionary biology, structural explanations are routine and unproblematic.

That said, it is not standard practice in biology to call structural explanations "structural explanations"; instead, examples of this kind of explanation are usually called "ultimate explanations," which are contrasted with "proximate explanations." The category of proximate explanation is broader than the category of individual or psychological explanation, but all individual and psychological explanations are proximate explanations.

Yet these are all semantic points that are of little consequence for the argument in this chapter. What is more important is that the ubiquity of structural explanations in evolutionary biology implies that there is a very simple way of "being biological" about the social behavior of humans. This stance starts with taking up the idea that structural explanations of human social behavior can be scientifically adequate explanations whether or not they refer to past cases of natural selection or contemporary social processes. Then, the proposal is to construct explanations for human behavior that mirror both the *types of* and the *relations between the types of explanations* as can be found in evolutionary biology. So, nonbiological explanations of human behavior can imitate explanations in biology, not by borrowing key biological concepts like "adaptation" and applying them to, for instance, traits and patterns of behavior in present-day humans, but instead by mirroring some of the deeper patterns in the explanations used in evolutionary biology by using, specifically, the same *types* of explanations in nonbiological contexts. And as we shall soon see in some detail, biologists keep ultimate explanations and proximate explanations conceptually and inferentially isolated from one another (Alcock and Sherman 1994)—so, one way of "mirroring" the relationship between structural and individual explanations as they are used in evolutionary biology is to recognize the scientific adequacy of structural explanations of human behavior and thereafter enforce the same pattern of isolation when constructing structural and individualistic explanations of human behavior. That, in a nutshell, is what it means to "be biological" from the point of view of social-turn moral psychology.

However, that way of "being biological" must be distinguished from a much more common approach that attempts to arrive at explanations of human behavior by inferring substantial psychological conclusions from evolutionary explanations. This alternative way of "being biological" about human behavior considers only the structural (ultimate) explanations that may have been true of human behaviors in the late Pleistocene and that refer to natural selection. It does not consider relevant any structural explanations of human behavior that refer to contemporary cultural, institutional, or political factors. Why? According to this style of "being biological" it is thought, mistakenly, that it is possible to reliably derive novel psychological conclusions from ultimate explanations, and where these conclusions almost always posit psychological mechanisms that are innate and nonmalleable. This provides impetus for the idea that social behaviors that occur in times and places more recent than the Pleistocene (the environment of evolutionary adaptation, or "EEA," for *Homo sapiens*) can be explained in a bottom-up fashion by treating the posited innate and nonmalleable psychological mechanisms as the most stable and common causes of the relevant patterns of contemporary social behavior. So, on this second approach to "being biological" about our social behavior, the use of structural explanations to understand human social behavior stops when the EEA for humans ends, while at the same time, when explaining patterns of behavior in the EEA, structural (ultimate) explanations are not kept fenced off from psychological (proximate) explanations, because specific examples of the latter are taken to be derivable from the former.

This chapter demonstrates why the latter pattern of explanation is unreliable, and therefore why attempting "being biological" about morality in the second way is a scientific mistake. It is almost never reliable to draw specific psychological conclusions from evolutionary explanations of patterns of human behavior in the EEA. A fortiori, even if we are given an extremely well-confirmed evolutionary explanation for why patterns of moral behavior were adaptive in the EEA, it is unreliable to accept any further claims about instincts or modules or learning routines or any other specific psychological mechanisms whose function may have been to cause the relevant patterns of behaviors. The correct conclusion really is that *no* nontrivial psychological conclusion can be inferred from evolutionary explanations of human behavior.

That general conclusion is important for a subtle but absolutely crucial reason. Evolutionary biology is among the most—if not the most—successful scientific disciplines of the last two hundred years. It is astonishingly unlikely that contemporary evolutionary theory could ever be replaced by a better comprehensive biological theory that is incompatible with the basics of the Darwinian approach. Because of this, it is important that judgments of scientific plausibility that are grounded in established evolutionary theory be taken very seriously. Why? Scientific hypotheses that are implausible in light of evolutionary theory are, for just that reason alone, unlikely to be true; and scientific hypotheses that are plausible in light of evolutionary theory are, again, for just that reason, likely to be true. Because of that, the argument that follows shows—and does so using only conceptual and methodological materials that are internal to evolutionary biology—that it is a mistake to draw psychological conclusions from ultimate explanations that describe adaptive patterns of human behavior. So, going in the other direction, scientific research projects that rest upon extracting specific psychological conclusions from evolutionary explanationsare in fact inconsilient with evolutionary biology, and for that reason, likely in the business of generating mostly false conclusions about human psychology.

Proximate and Ultimate Explanations

Let us now take a closer look at the distinction between proximate and ultimate explanations. Here is John Alcock's explanation of this distinction: "In the jargon of biologists, studies of how cellular mechanisms and systems-operating rules influence behaviour are classified as *proximate* research, which examines the immediate causes of the traits of interest. In contrast, questions about the adaptive (reproductive) value of behaviour are labeled *ultimate* questions, not because they are more important than proximate ones but because they are different, dealing with the long-term historical causes of the special abilities of species" (Alcock 2001, 12–13; emphasis in original). Thomas Scott-Phillips, Thomas Dickins, and Stuart West write: "The difference between proximate and ultimate explanations of behavior is central to evolutionary explanation ... ultimate explanations address evolutionary function (the 'why' question), and proximate explanations address the way in which that functionality is achieved (the 'how' question). Another way to think about this distinction is to say that proximate mechanisms are behavior generators, whereas ultimate functions explain why those behaviors are favored" (Scott-Phillips, Dickins, and West 2011, 38) And here is Mary Jane West-Eberhard's definition of the very same distinction: "The answer to 'why' an organism is, or behaves in, a certain way can be answered either in terms of mechanisms (proximate causes) or in terms of selection and evolution (termed 'ultimate causes'[...]). This distinction is designed to prevent confusion between levels of explanation in biology" (West-Eberhard 2003, 10) Recall from chapter 1 the idea of conceptual framework pluralism: the distinction between proximate and ultimate explanations is one of the specific ways that this pluralism is instantiated in the practice of evolutionary biology.

Scott-Phillips, Dickins, and West's use of the concept of a behavior generator here is important for further understanding the interaction between ultimate and proximate explanations. This is because a trait that makes no difference to behavior can be neutral with respect to natural selection. Psychological traits can therefore be the subject of ultimate explanations only indirectly; it is only insofar as they function as behavior generators that psychological traits are able to make the kinds of differences that induce natural selection. It is the *behavioral* part of the concept of the behavioral generator that should be stressed. To see why that is the case, imagine that we have two hypotheses about different psychological mechanisms, and that both of these mechanisms motivate exactly the same pattern of adaptive behavior in exactly the same circumstances. Knowing that natural selection would have favored the relevant behavior does not provide us with any insight into which of our two hypotheses are true; more generally, knowing the correct ultimate explanation for a trait (an explanation that demonstrates that the trait is an adaptation) will not help decide which of the differing proximate hypotheses is correct. And the same is true for ultimate explanations: knowing the true proximate explanation for a trait will almost never help you identify the correct ultimate explanation. That is why Scott-Phillips and colleagues conclude, "[Answers] to any [proximate] questions does not commit us to any particular answer to the other [ultimate] questions. They are not opposite ends of a continuum and we should not choose between them. On the contrary, they are distinct from one another and complementary. To completely understand behavior, we must obtain both" (Scott-Phillips, Dickins, and West 2011, 38).

It is sometimes an entailment of explicit definitions of the proximate–ultimate distinction that the distinction applies only to adaptations. For example, here is a definition from Andy Gardner:

> Proximate explanations concern the mechanism by which the adaptation operates, including environmental triggers, ontogeny and the physics and chemistry underlying the final trait. That is, they provide an account of how genetic information is translated into a phenotype that manifests design. Ultimate explanations concern the adaptive rationale of this design. That is, they provide an account of how the phenotype is translated into the organism's fitness. Proximate explanations have been described as the domain of the functional biologist, and ultimate explanations, the domain of the evolutionary biologist. However, healthy biological science seeks an interplay between the two, leading to a fuller understanding of biological adaptation. (Gardner 2013, 787–788)

Gardner's definition is very the similar to the previous three, but for various technical reasons, definitions like Scott-Phillips and coauthors' should be preferred. There are proximate explanations for many traits that are not adaptations and yet are subject to ultimate explanations. In the simplest case, there are "behavior generators" in many species that cause behaviors that are themselves neutral with respect to any form of natural selection. These behaviors are neither adaptive nor maladaptive. However, to determine that any of these behaviors in fact are neutral with respect to selection is ipso facto to provide an ultimate explanation of the behaviors. This is not just a philosophical concern—many novel mutations are neutral traits (Nei, Suzuki, and Nozawa 2010). So, it is a mistake to define ultimate explanations as applying only to adaptations.

It is hard to overstate how important the proximate-ultimate distinction is to evolutionary biology. The distinction explains why, for example, the conceptual vocabulary of neurophysiology cannot be inferred from that of behavioral ecology, and also why both of these vocabularies can be combined to generate explanations of, for example, the relationship between scent and sexual attractiveness in both humans (Thornhill et al. 2003) and nonhumans (Martín and López 2006). Even biologists who are skeptical of the proximate-ultimate distinction agree that the "vast majority of behavioral hypotheses" nevertheless fit the distinction (Dewsbury 1994, 64). We might even say that nothing in the recent history of biological science makes sense except in the light of the proximate-ultimate distinction.

Yet there is a logical property of the proximate-ultimate distinction that is rarely discussed. To bring out the relevant property, I need to introduce a distinction between theories that are plausible on the basis of prior scientific research but are nevertheless not yet confirmed, and then, the smaller number of theories that are able to transition from that category into the category of confirmed scientific theories. I will call theories in the first category "scientifically plausible" theories, and theories in the second category "confirmed theories." Given this, the important property is that the relationship between scientifically plausible proximate and scientifically plausible ultimate explanations is many-to-many: for any particular trait or behavior, past scientific research will almost always provide support for multiple different proximate and multiple different ultimate explanations.

Here is how John Alcock and Paul Sherman describe the many-to-many relationship and the role it plays in organizing biological research:

Some persons derive satisfaction from teasing apart the genetic, neuronal, developmental, or cognitive proximate mechanisms that enable an animal to carry out a particular response. Others of an ultimate bent prefer to investigate the evolutionary history of the mechanism or the current selective consequences of the behavior the mechanism controls. Some individuals are afflicted by the kind of research chauvinism that leads them to think that whatever they are interested in must be the most important of all topics. But this attitude does not result from recognizing that proximate causes differ from ultimate ones, or that there are different levels of analysis within each category. Moreover, the idea that all researchers should integrate multiple levels of analysis is both impractical and unnecessary, provided that there is open communication between workers. We believe there is a general awareness that it is helpful to use evolutionary theory to inform proximate research, just as it can be productive to use knowledge about internal mechanisms and ontogenies to raise interesting ultimate questions. [...] The whole point of making the proximate-ultimate distinction was ... and still is ... pertinent, namely to eliminate the senseless arguments that arise from confusing these two logically distinct levels of analysis. (Alcock and Sherman 1994, 61)

Let me provide an example to illustrate Alcock and Sherman's methodological conclusions. In an earlier field study, Sherman investigated the ultimate function of warning calls in Belding's ground squirrels (*Spermophilus beldingi*). Trained raptors flew over natural colonies of these squirrels, and Sherman analyzed the effects of warning calls produced by the squirrels (Sherman 1985). Sherman's initial hypotheses offered two ultimate explanations: the warning calls are nepotistic (they primarily benefit genetically related individuals) or they are self-preserving (they benefit the squirrel that produces the sound). Sherman's data showed that the specific warning calls that Belding's ground squirrels use for aerial predators (a single whistle) helped the squirrel that produced the sound escape as much as it helped other squirrels. However, the warning call used for terrestrial predators (a multinote trill) did not confer this advantage: squirrels who warned of terrestrial predators were more likely to be killed by the predator than those squirrels who warned of aerial predators. Sherman concludes, "while trills are expressions of nepotism, whistles involve self-preservation."

Two observations are relevant. First, prior to examining his data, Sherman had two scientifically plausible ultimate explanations for the warning calls, and his data helped him determine the specific kinds of calls to which these explanations apply. More importantly, Sherman did not reach his conclusion about the ultimate functions of the warning calls by analyzing the proximate facts about squirrel vocal physiology, or the computations in the visual system that underlie the ability to detect aerial predators against both cloudy and clear skies, or investigating any of the hundreds of other proximate questions that can be asked about how individual squirrels are able to recognize airborne raptors. These are all interesting and important areas of research, but confirmed scientific theories about these various proximate phenomena were not needed in order for Sherman to make reliable inferences about the probable ultimate explanation for warning calls. Indeed, Sherman's methodology is both neutral with respect to the truth of different proximate explanations that might be true of the ground squirrel's visual, auditory, vocal, and cognitive systems, and more important, his data does not help determine which of these many scientifically plausible proximate explanations are true. This is an illustration, then, of the many-to-many relationship that holds between scientifically plausible proximate and ultimate explanations. Producing evidence that helps determine which member of a set of scientific plausible ultimate explanations is true does not yield any insight into which scientifically plausible proximate explanations are likely to be true, and vice versa.

It is easy to find further examples of exactly the same many-to-many relationship between scientifically plausible proximate and ultimate explanations. For example, there is a wider literature that studies the ultimate function of warning calls in rodents (Blumstein and Armitage 1997; Blumstein et al. 1997; Shriner 1998). For a striking illustration of just how complex the relationship between proximate and ultimate explanations can be, take the literature, more than a century old, on the ultimate functions of coloration (Jones 1932; Thompson 1960; Smith and Harper 1995; Krajíček et al. 2014). Of particular interest here are the gastropods. Different ultimate explanations have been proposed for the morphological traits for numerous gastropod species related to coloration (Cooke et al. 2014; Ornelas-Gatdula, Dupont, and Valdés 2011). However, the sheer number of gastropod species should give pause to anyone who thinks that it is easy to infer proximate explanations from ultimate explanations of even relatively common traits like, for instance, shape and brightness. Recent (albeit rough) estimates place the number of living gastropod species at over fifty thousand (Bouchet and Rocroi 2005). The point is that for each species there will be at least one proximate explanation that describes the developmental pathway by which traits apparently common to several species are produced, and there may also be an ultimate explanation for such traits that fits different species across a clade or even the class. Yet confirming an ultimate explanation of, for instance, how coiled shells with brightly colored red specks makes a selective difference to the different species with those traits would nevertheless provide no insight into the different proximate processes by which shells develop across all of the species that have evolved this trait.

So there is an important methodological reason for explicitly conceptualizing the proximate and ultimate explanation as a many-to-many relation. Failing to understand that the relationship between these two types of explanation is many-to-many risks rendering the distinction inconsistent with the diversity of mechanisms and processes at both the ultimate and proximate levels in the biological world. Treating the relationship as a one-to-one relationship is, in effect, to fail to appreciate the complexity of the biological world.

It is time now for me to introduce the definitions that I will be using. I want to depart somewhat from standard practice by including a third component of the distinction, which I propose be called Mayr's lemma, if only because the proximate-ultimate distinction was introduced to biology and philosophy by Ernst Mayr in the course of his analysis of Tinbergen's earlier four-questions distinction (Mayr 1961; Tinbergen 1963).

(PE) Proximate explanations refer to proximate causes and mechanisms. These causes and mechanisms are mostly endogenous to the body of particular organisms, and they occur within the lifetime of the organism. Proximate mechanisms are usually genetic, biochemical, physiological, neurophysiological, computational, or cognitive entities.

(UE) Ultimate explanations refer to ultimate causes and mechanisms. These causes and mechanisms are those that occur exogenously to the body of particular organisms, they affect more than just individual organisms, they can occur over a long period of time, extending beyond the life of any particular organism, and they almost always refer to structural properties and constraints of the environment in which groups or populations of organisms live. Ultimate causes and mechanisms are population-level phenomena such as natural selection, migration, and many other metaphysically similar processes.

(Mayr's lemma) Scientifically plausible ultimate explanations and scientifically plausible proximate explanations stand in a many-to-many relationship. It will not generally be possible to infer which proximate explanation of some pattern of behavior is true from a confirmed ultimate explanation of the pattern; and it will not generally be possible to infer which ultimate explanation of some pattern of behavior is true from a confirmed proximate explanation of the behavior.

As the earlier quotes about the proximate-ultimate distinction show, Mayr's lemma is not included as an explicit component of the distinction between proximate and ultimate explanations. But it deserves to be, because it is crucial to understanding how to apply the distinction in sound biological practice.

I will sometimes say that different examples of scientific reasoning are violations of Mayr's lemma in this and the following chapters. What I mean by that claim is that the example in question has presupposed, wrongly, that the relationship between scientifically plausible ultimate (or structural) and scientifically plausible proximate (or psychological) explanations is *not* many-to-many. The most common form of this fallacy is to mistake the logical compatibility of one scientifically plausible ultimate explanation with another scientifically plausible proximate explanation as evidence that substantially increases the degree of confirmation of both hypotheses. This is a mistake that is easy to see if we change our focus only slightly. Observing that a scientifically plausible hypothesis in political geography about a pattern of human behavior is logically compatible with a scientifically plausible hypothesis in developmental psychology about the same pattern does not provide us with reason to think that both hypotheses are, just and only for that reason, closer to being confirmed. Yet exactly that pattern of inference is extremely common in evolutionary psychology (Machery, forthcoming) and also, as we shall soon see, has influenced some moral psychologists.

That said, Mayr's lemma should not be understood as implying that ultimate and proximate explanations stand in an any-to-any relationship. In the biological world there is no clean and clear metaphysical separation between ultimate and proximate phenomena. Ultimate and proximate processes are always interacting, and increases in our scientific understanding of these interactions can lead to different estimates about which novel ultimate and proximate hypotheses are, at some particular time, scientifically plausible. Mayr's lemma says only that the relationship between scientifically plausible proximate and ultimate explanations is many-to-many, and not either one-to-one or any-to-any.

Explanatory versus Predictive Uses of Ultimate Explanations

There is one more technical distinction that we must learn before we can fully understand what it means to violate Mayr's lemma. We need to distinguish between *predictive* and *explanatory* uses of ultimate explanations.

Here are the details. Suppose that we have observed that a trait—like morning sickness in humans—is characteristic of a species. We know, for instance, that women who experience morning sickness are less likely to miscarry, that morning sickness peaks during weeks six to eighteen of pregnancy, which is when the embryonic tissue is most susceptible to chemicals that can cause deformations, and that societies with lower levels of morning sickness are frequently societies with higher corn than meat consumption. Such facts about morning sickness have led a number of scientists to constructive speculative ultimate explanations for morning sickness (Flaxman and Sherman 2000). The motivation here is the view that nausea and vomiting during pregnancy may confer a comparative advantage in terms of fitness, if only to the extent that having the mechanisms makes a marginal difference to the viability of a fetus. But all the same, there is absolutely no scientific debate over the existence of morning sickness. And that is the key point. Given that we have already identified some specific proximate mechanism (in this case, morning sickness), we can begin to evaluate different ultimate explanations for the same mechanism (or the behavior it produces) in order to determine, for instance, if the mechanisms (or their behavioral effects) are an adaptation. Call this use of ultimate explanations *explanatory* uses of ultimate explanations. Explanatory ultimate explanations are *not* used to derive novel psychological conclusions—they are, instead, used to determine if there is a credible evolutionary story to tell about some particular proximate mechanism, the existence of which has been settled by a line of independent scientific research. Or, to generalize now, suppose that we have observed that most members of a species have a particular trait T. Given that, we can try to find an ultimate explanation of T, and if we are able to confirm an ultimate explanation of T that refers to some form of natural selection, we have learned that T is an adaptation. This is an *explanatory* use of an ultimate explanation. This ultimate explanation does not tell us *that* T exists; it only tells us *why* T exists.

Explanatory uses of ultimate explanations are to be contrasted with *predictive* uses of ultimate explanations. It is the predictive use of ultimate explanations that routinely leads to violations of Mayr's lemma, and is therefore unlikely to be reliable. A predictive use of an ultimate explanation begins by formulating a scientifically plausible ultimate explanation for some pattern of behavior B, and then *predicts* the existence of a specific, discrete proximate mechanism on nothing more than the grounds that the proximate mechanism is a possible behavior generator for B. Crucially, other scientifically plausible proximate explanations for B are ignored—because, technically, to use E to predict B is to deduce B from E and its auxiliary hypotheses, and so if E is probably true, then B is also probably true. However, Mayr's lemma implies that you cannot derive predictions about scientifically plausible proximate explanations from one single scientifically plausible ultimate explanation, because the Lemma says that there will be multiple different scientifically plausible explanations that are compatible with the relevant ultimate explanation. So, predictive uses of ultimate explanations—which, again, are *predicting* the existence of a *single* and otherwise unknown proximate mechanism on nothing more than the grounds that a scientifically plausible ultimate explanation shows that the hypothetical mechanism *could* have caused a putatively adaptive behavior in the EEA—violate Mayr's lemma. Mayr's lemma says that there will be *multiple* different scientifically plausible proximate explanations for B; but the predictive inference assumes that there is just one proximate explanation compatible with the ultimate explanation for B.

Suppose that we have identified both the EEA for a particular pattern of behavior B, and that we also know that in this environment natural selection favored B. That is the same as saying that we have a confirmed ultimate explanation for B, and that this explanation tells us that B is an adaptation. However, this does not help us identify which proximate mechanisms are causing the relevant behavior. As we know by now, that is because Mayr's lemma tells us that there will be a range of scientifically plausible proximate explanations that are compatible with our confirmed explanation. It is helpful to see that these scientifically plausible—but not yet confirmed—proximate explanations are *behaviorally equivalent* with one another: each proximate explanation predicts that B will occur. And so, the point of this hypothetical example is that, even if we assume that we have a confirmed ultimate explanation for some behavior B, which therefore tells us that B is an adaptation, we still cannot draw any conclusions about which scientifically plausible proximate explanations for B are true.

Evolution, Morality, and Violations of Mayr's Lemma

Morality is surely some a kind of social behavior, and we clearly have evidence of its existence from fields other than evolutionary biology. There can be no objection, thus, to explanatory uses of ultimate explanations that seek to understand how morality, or something very much like it, may have evolved. For example, see Nowak 2006 for a brief survey of different kinds of ultimate explanation that can be put to *explanatory* (not *predictive*) use in order to account for the evolution of morality. These explanatory uses of ultimate explanations are psychologically—and neurophysiologically, developmentally, and so on through the various proximate levels of description—neutral; they refer, instead, only to the population-level and therefore structural processes that may have functioned to conserve moral behavior. These ultimate explanations have no direct implications for which theories in moral psychology are likely to be true or false, and confirming any of these ultimate explanations will therefore leave any number of important proximate questions unanswered.

But that is not the case for predictive uses of ultimate explanations that are applied to the evolutionary history of morality. They often may seem like a viable way of bringing evolutionary biology and morality together—until, that is, the implications of Mayr's lemma are properly understood. Instead, and somewhat ironically, predicative uses of ultimate explanations applied to the history of morality lead to inconsilience between evolutionary biology and moral psychology. The argument, in its simplest form, is this. Either evolutionary biology is or is not reliable. If it is reliable, then part of the reason why it is reliable is that the proximate-ultimate distinction is, when applied to scientifically plausible theories, a many-to-many relation. This is part of what allows conceptual framework pluralism to be true of evolutionary biology. So, if evolutionary biology is reliable, predictive uses of ultimate explanations are unreliable. Going in the other direction, however, if predictive uses of ultimate explanations are reliable, then certain deep methodological principles baked into contemporary evolutionary biology are false and evolutionary biology is, despite all evidence to the contrary, unreliable. In either case, then, there is inconsilience between present practice in biology and predictive uses of ultimate explanations, and it almost goes without saying that the weight of scientific evidence entails that predictive uses of ultimate explanations should be abandoned.

That said, this section provides several examples of predicative uses of ultimate explanations, and this evidence is part of the motivation for the remaining chapters of the book—for these examples show that there really is good reason to explore alternative ways of bringing evolutionary biology and moral psychology together. In other words, it is not hard at all to find scientists and philosophers who think that the way to bring evolutionary biology and moral psychology together is to rely on different versions of predictive uses of ultimate explanations.

Jason Mckenzie Alexander's aptly named book *The Structural Evolution of Morality* relies almost entirely upon explanatory uses of ultimate explanations to advance its argument. McKenzie Alexander uses a family of replicator equations to show that several different patterns of cooperation are evolutionarily stable. His explanation of what these models reveal about moral behavior is worth understanding in detail. He writes:

> Morality comprises a set of heuristics that govern behavior in interdependent decision contexts that work to the benefit of each person in society when followed. That is, acting in accordance with those heuristics serves to maximize satisfaction of each person's preferences, as much as is possible, given the constraints placed upon satisfaction of our presences by the fact that many people's preferences are in conflict. The reason why such heuristics (moral norms) work in this way derives from the social structure of interdependent decision problems whose choices they regulate. The fact that most people (perhaps all people) do *not* know why or how the heuristics work is of no real consequence: most people cannot explain, or justify, the functioning of the heuristic they use to catch fly balls.
>
> This book offers a preliminary argument for the claim that many of the norms governing behavior, in particular, certain primitive moral principles, can be understood as general heuristics whose adoption ensures that an individual will generally *do better* if they are followed that if they are not followed. These social norms exist as a culturally evolved response to repeated interdependent decision problems that occur in a socially structured environment. ... The claim that people will generally do better following moral norms than not should not be confused with the similar-sounding claim that people have deliberately *chosen* to adopt the norms *on the grounds that* acting in accordance with them serves to maximize their individual expected "utility." (McKenzie Alexander 2010, 23)

McKenzie Alexander is clearly reluctant to draw any psychological conclusions from the structural dynamics his models characterize. To see this, note that the first paragraph in the preceding quote is ambiguous in a way that the second paragraph is not. Since it is possible for someone to be following a heuristic or a rule without having an exact representation of that rule in her head, there is no inference from the fact that, behaviorally, people are implementing a rule or acting in a way that is maximizing their utility while also minimizing their risk or uncertainty, to any particular psychological conclusions about the psychological cause of this behavior. By the next paragraph, McKenzie Alexander resolves this ambiguity: the structural conclusion that people will do better implementing moral norms (that is, the evolutionary stable principles identified by his models) has no nontrivial implications about the properties of the proximate mechanisms that either would or did cause the relevant norm-following behavior. The models that he defines are psychologically neutral, in the specific sense that, for each, it is possible to construct different behaviorally equivalent proximate explanations for whatever patterns of behavior are shown to be stable or optimal by the models. It follows that the models cannot be used to determine which, if any, of these proximate explanations are true. Accordingly, McKenzie Alexander reaches the following conclusion:

> I must admit skepticism as to whether evolutionary game theory *on its own* is capable of providing thick descriptions of moral behavior. The reason behind this skepticism is simply that fine-grained accounts of individual psychology are not the sort of thing which falls within the domain of the theory. In the original biological setting, evolutionary game theory analyzed problems in which frequency-dependent fitness introduced a strategic element in evolution. In the cultural setting, evolutionary game theory examines repeated interdependent decision problems played in populations of boundedly rational individual. In both settings the theory tracks only chances in the frequencies and distributions of strategies and, perhaps, other relevant properties. Under either interpretation, evolutionary game theory addresses only the *strategic* aspect of the superstructure, not the *psychological* aspect. (McKenzie Alexander 2010, 270)

I suggest that these remarks show that McKenzie Alexander is aware that it is standard practice when thinking about morality from an evolutionary perspective to draw "think"—meaning, *psychological*—conclusions from structural explanations. To his credit, McKenzie Alexander recognizes that this is not possible.

That is not true, unfortunately, for our remaining examples. These alternatives follow the lead of E. O. Wilson and his treatment of human morality in his famous *Sociobiology: The New Synthesis*. Wilson writes:

> Scientists and humanists should consider together the possibility that the time has come for ethics to be removed temporarily from the hands of the philosophers and biologicized. The subject at present consists of several oddly disjunct

conceptualizations. The first is *ethical intuitionism*, the belief that the mind has a direct awareness of true right and right that it can formalize by logic and translate into rules of social action. ... The Achilles heel of the intuitionist position is that it relies on the emotive judgment of the brain as though that organ must be treated as a black box. (E. O. Wilson 2000, 562)

Wilson then conjectures that science can open this black box, discover its genetic programming, and thus explain the content of people's intuitions while at the same time debunk any claim these intuitions have to either genuine normative standing or epistemic reliability. And to be clear, there is nothing wrong with this hypothesis, just so long as it is treated as a scientifically plausible proximate hypothesis, and therefore as one of many different proximate explanations that one might produce for, inter alia, the phenomena of individual moral judgment. But that is not Wilson's intent. Instead, he argues that scientific inquiry into the dynamics of past selection on both humans and yet more distant ancestral species will reveal why the genes for moral judgment were conserved, and consequently, what constraints must be hard-wired into human moral cognition. So, Wilson falls afoul of Mayr's lemma because he only seriously considers a single proximate hypothesis, namely, that moral judgments can *only* be caused by some kind of innate and nonmalleable moral mechanism.

For a more recent example of the same mistake, we can turn to Martin Daly and Margo Wilson's theory of homicide. They write:

If moral sensibilities have evolved by the utilitarian criteria of natural selection, it follows that acts which appear to be motivated by deference to interests other than the actor's own have routinely rebounded to the benefit of the moral actor. [...] The ostensibly alternative philosophical positions of retributivism and utilitarianism can be considered to constitute different levels of analysis. From the perspective of evolutionary psychology, the retributivist may be said to consult evolved moral/cognitive/emotional mechanisms to apprehend what *feels just*. Utilitarian considerations are then relevant to the task of explaining why such mental mechanisms have evolved. [...] We punish, *should* punish, *must* punish ... from the perspective of evolutionary psychology, this almost mystical and seemingly irreducible sort of moral imperative is the output of a mental mechanism with a straightforward adaptive function: to reckon justice and administer punishment by a calculus which ensures that violators reap no advantage from their misdeeds. (Daly and Wilson 1988, 255–256)

But again, even if we grant that natural selection is and was "utilitarian," Mayr's lemma tells us that there will be multiple different, and behaviorally-equivalent, proximate mechanisms that are compatible with whatever the specific dynamics of utilitarian natural selection are. Daly and Wilson are wrong that "it follows" from even these assumptions that there is a psychological mechanism that produces thoughts of punishment about, for instance, free riders.

Then there are any number of treatments of moral intuition by social psychologists and moral philosophers both that conceptualized moral judgments as the output of some innate, evolved moral module (Haidt 2001; Greene and Haidt 2002; M. D. Hauser 2006; M. Hauser et al. 2007; Cushman and Greene 2012). Indeed, it is no accident that Haidt collects these different projects together in a review article titled *The New Synthesis in Moral Psychology* (Haidt 2007). What all of these moral sense theories have in common is that moral judgment is assumed to be generated by some kind of nonmalleable moral instinct, where the evidence for the existence of this instinct is not psychological or neurological in nature but rather evolutionary: the existence of the relevant mechanisms is simply taken to "follow" from descriptions of distinct processes of natural selection. This is obviously a violation of Mayr's lemma. Again, one can grant that these processes did in fact occur and still see that no substantial psychological (or any other kind of proximate explanation) is thereby implied.

Still, let me provide two detailed examples of this moral instinct–positing theories of morality so we can see some of the finer details of this fallacious pattern of inference. The first of these is from Daniel Krebs, who begins one of the more sophisticated efforts to explain the evolution of morality with these words:

> At an intuitive level, everyone knows what a sense of morality is. After all, we all possess one. It is a mental phenomenon that consists in thoughts and feelings about rights and duties, good and bad character traits (virtues and vices), and right and wrong motives and behaviors. It contains feelings of entitlement and obligation, as well as such evaluative feelings as pride, respect, gratitude, forgiveness, guilt, shame, disgust, and righteous indignation. It contains standards of conduct; conceptions of distributive, contractual, restorative, retributive, and procedural justice; and thoughts and feelings about purity, sanctity, and sin. ... Many people's conceptions of morality are embedded in religious beliefs. (Krebs 2008, 150)

The phrase "embedded in religious beliefs" is telling, as one can reasonably expect that personal religious affiliation and upbringing can provide an independent proximate explanation of at least some normative beliefs and feelings. But Krebs aims instead to find an evolutionary explanation of all forms of normativity, not just morality, and he does this by developing a sophisticated theory of how instinctive moral intuitions constitute this "sense of morality." He continues thus: "the proximate mechanisms that give rise to prosocial behaviors produce affective experiences that contribute to a primitive sense of morality in humans and, perhaps, in other primates. [...] When the evolved mechanisms that induce people to behave in prosocial ways are activated, they produce psychological states that people experience as fear, awe, righteous indignation, disgust, guilt, shame, gratitude, forgiveness, contrition, love, sympathy, empathy, loyalty, or a sense of solidarity" (Krebs 2008, 149).

So, what evidence is there that these complex proximate mechanisms exist? Krebs presents an updated version of Darwin's own theory of the evolution of morality that treats morality as encompassing various different patterns of cooperation. However, Krebs's deeper question is not whether or not the relevant patterns of behavior could or could not have evolved, but instead "how mental mechanisms that dispose individuals to help other members of their groups could evolve" (Krebs 2008, 151). In so doing, Krebs begins to conflate what are otherwise both conceptually and scientifically distinct ultimate and proximate explanations. For example, "Fundamental social dilemmas are exacerbated by natural selection. If those who are naturally disposed to behave selfishly contribute more than their share of offspring to future generations, the proportion of selfish members in a group will increase. As the number of selfish members increases, the benefits of selfishness will decrease, because fewer and fewer members will be disposed to behave cooperatively and, therefore, there will be progressively fewer cooperators to exploit" (Krebs 2008, 151–152). This mixing together of proximate and ultimate conditions leads Krebs to reason, fallaciously, according to what might be called an evolutionary ladder. A single proximate hypothesis is paired with a single ultimate hypothesis, and the logical compatibility of these explanations is treated as evidence of their mutual truth. The initial pair is then used as the basis for inferring another ultimate-proximate pair, which then, when combined with the first pair, leads to the inference of a third pair, and so on until Krebs concludes that he has explained all of the properties of the human moral sense. However, and at most, this reasoning shows that a network of different proximate and ultimate explanations are logically consistent with one another—what is required, furthermore, to know that any of these explanations is true is, first, to add to the set of hypotheses under consideration *all* scientifically plausible proximate and ultimate explanations for the various different patterns of social behavior, and then to find empirical evidence that can be used, ideally, to falsify all but a small number of the total set of scientifically plausible proximate and ultimate explanations.

Richard Joyce's recent *The Evolution of Morality* follows a similar tack. Joyce begins by establishing that some fitness-sacrificing ("helpful") behaviors could have evolved. From this, he writes, "The question in which I am interested is 'What proximate mechanisms might be favored by natural selection in order to regulate this helpful behavior?' One possible answer, which I think is correct, is 'Altruism' ... [which is] the trait of acting from altruistic motives ... Another possible answer, which I also think is correct, is 'Morality.' In order to make an organism successfully helpful, natural selection may favor the trait of making moral judgments" (Joyce 2007, 255). Joyce proceeds to offer a sophisticated ultimate explanation of how helpful behavior can have evolved. But the only proximate explanation that he seriously considers is that the neurological mechanisms implicated, inter alia, in maternal care for newborns are "fiddled with" by evolution to eventually produce a moral sense, although only after other environmental factors unique to human life, like language, have also evolved. Joyce considers a number of different formulations of a moral sense theory—but the tenor of his analysis is accurately summarized by Joyce's sense of his key questions, which, by the middle of his book, are: "Why would a moral sense evolve by natural selection ... [and] How did a moral sense evolve?—What did natural selection do to the human brain to enable moral judgment?" (Joyce 2007, 108). Joyce is sensitive to one of the problems with this explanatory approach. He writes that answering these questions "will find us with a big fat just-so story about the evolution of morality on the table. Since I am averse to leaving matters at that, my third task ... is to conclude with a review of some empirical evidence in support of the story, thus turning it into a respectable scientific hypothesis" (Joyce 2007, 108). This in turn becomes an effort to show that the hypothesis that "morality (by which I here mean *the tendency to make moral judgments*) exists in all human societies we have heard of" is testable (Joyce 2007, 134; emphasis in original). We can move past the conflation of proximate and ultimate definitions of morality here—what Joyce means by "testable" is that there is empirical evidence that, if false, would substantially reduce the plausibility of both his ultimate hypothesis (morality exists) and his proximate hypothesis (there is a universal psychological tendency to make moral judgments). This evidence is an assortment of anthropological claims about the cultural ubiquity of morality, and some evidence from developmental psychology indicating that children have the ability to use some basic moral concepts.

That's a sophisticated argument. But the problem with it is simple. Joyce unfortunately overlooks one of the crucial lessons of Gould and Lewontin's critique of just-so stories in biology (Gould and Lewontin 1979), and because of this misunderstands both the kind of evidence that is needed to support his just-so story, and the power of these stories to confirm proximate hypotheses like his moral sense theory. Gould and Lewontin have two methodological lessons, the first of which is widely understood, and the second of which is almost always completely ignored. The first lesson is that, when considering known proximate traits, it is important to begin with the question of *whether* the trait is an adaptation before turning to an investigation of *how* the trait could have increased the fitness of its bearer. That is the basis of Gould and Lewontin's critique of the adaptationist program in evolutionary biology: do not assume a priori that every trait is an adaptation. The second lesson is easy to grasp once Mayr's lemma is understood. Gould and Lewontin also argue that the *how* question is not answered when only a *single* scientifically plausible ultimate explanation for the trait is produced. Why? Just as Mayr's lemma says, there will be other plausible ultimate explanations that can also be produced, some of which may not refer to kinds of natural selection. Simply being able to produce *a single* empirically compelling just-so stories—that is, scientifically plausible ultimate explanations that show how a trait could have been an adaptation—does not stand in for collecting the total empirical evidence needed to begin to rule out all but one of these ultimate explanations.

So, Joyce's mistake is not that he failed to recognize that ultimate explanations require empirical evidence; rather, it's that he failed to recognize that the relevant empirical evidence, if it is able to confirm any scientifically plausible ultimate explanation, must be able to decide among *all* initially scientifically plausible ultimate explanations. The mistake, then, pertains to how much empirical evidence is needed to show that some scientifically plausible ultimate explanation is also true, and not just that some speculative ultimate explanation can possibly be supported by empirical evidence and therefore counts as a scientifically plausible ultimate explanation. The relevant empirical evidence does not aim to show that a just-so story is not really a just-so story; the right kind of empirical evidence is the evidence that shows that all but one of the initially plausible just-so stories really aren't so, and therefore that the one remaining just-so story is, probably, the one true really-so story.

Conclusion

What are some of the philosophical implications of this critique? First, there is no reason to assume, *even if morality (as a form of social behavior) evolved by natural selection*, the existence of a set of psychological mechanisms that caused the relevant adaptive moral behavior in the EEA and that also caused an important number of contemporary social or moral behaviors, or both. But more significantly, it also follows that, if we make exactly the same assumption, there is no reason to believe that the category *contemporary moral behavior* is a subset of the larger category *adaptive social behavior*. Why? It would follow that the latter category included the former if, for instance, we knew that the very same innate and nonmalleable psychological mechanisms that induced moral behavior in the EEA were also the *only* cause of moral behavior in contemporary society. But there is no evidence for this claim—and the claim is doubly implausible once we account for efforts to assess the interactions between developmentally endogenous proximate mechanisms and cultural factors in the evolution of sociability and cooperation (Boyd and Richerson 2009; Jablonka, Lamb, and Zeligowski 2014, chap. 8), as work in this area makes very clear the basic fact that cultural variation is a constantly necessary source of novel behaviors, even when some of these behaviors are eventually canalized because of their adaptive consequences. It is both a philosophical and a scientific mistake, then, to assume a priori that the potential diversity and range of behaviors that can fall into the category *contemporary moral behavior* are roughly proportional to the diversity and range of cooperative behaviors that would have been adaptive in the EEA.

A further sociological lesson about patterns of scientific reasoning seems to be that it is surprisingly easy be mistaken about how difficult it is to show that a pair of—or small sets of—ultimate and proximate explanations for some particular human behavior are both confirmed. The evidence necessary for such an achievement is evidence that can decide between *all* scientifically plausible proximate and ultimate explanations for some pattern of behavior, and not just evidence that is consistent with each member of the relevant pair.

But neither of those two lessons are the most important to draw from this chapter. What is important is that we now have a good understanding of a way *not* to bring together evolutionary biology and moral psychology. While biological reasoning may be able to provide us with credible ultimate explanation of moral behavior—or, at least, certain generic patterns of cooperation—these explanations will not yield any specific insights into how the mind implements moral cognition and how it motivates moral behavior. Again, the technical reason for this is that there will be multiple different *behaviorally equivalent,* scientifically plausible proximate explanations that are compatible with the relevant ultimate explanation, and knowing only that a pattern of moral behavior is adaptive will provide no guidance about which, if any, of these proximate explanations is true.

So, how do we decide between the explanations in moral psychology? And how do we do that work in a way that is not just methodologically neutral with respect to evolutionary biology, but instead also establishes a meaningful degree of consilience between the field and moral psychology? The position I develop over the remaining chapters of this book is captured fairly well by the following package of ideas: (1) moral psychology and ethics should "imitate" the inferential practice of evolutionary biology by keeping structural and individual explanations inferentially and conceptually separate, because scientifically plausible explanations of these types, even when they are applied to the social behavior of humans, should be understood to stand in a many-to-many relationship; (2) moral psychologists should take seriously those structural explanations that describe actual existing patterns of ethical behavior in different social contexts (this is the "social turn"); and (3) one of the core questions for moral psychology becomes determining which psychological explanation of the relevant ethical behavior is true.

Indeed, let me conclude by putting forward a brief argument for the idea that Mayr's lemma is true of the social behavior of humans, which is a conclusion that helps motivate (1) in the preceding paragraph. There are two parts to the argument. First, conceptually, it is important to understand (as noted at the beginning of this chapter) that ultimate explanations are a type of structural explanation, and individualistic or psychological explanations are a type of proximate explanation. Now, it would be a mistake to take these semantic facts as fixed a priori truths about how certain high-level concepts must be defined, because the relevant semantic facts have been established by induction over different histories of scientific inquiry; these histories have produced different first-order conceptual vocabularies that have proved useful for talking about, explaining, and sometimes even predicting the behavior of natural phenomena at different levels of organization. We have not yet discovered a single, unified conceptual vocabulary that is apt for describing, for instance, the full development process by which a zygote turns into a walking, talking scientist; neither have we found a single unified conceptual vocabulary that is apt for describing, for example, how democracies are formed and that also makes reliable predictions about how to permanently preserve any democracy in any society whatsoever. In respect of only these otherwise banal facts about the history of inquiry, the social and the behavioral sciences are no different than the biological sciences; Wimsatt was right to lump these sciences together by virtue of how well they respectively exemplify conceptual framework pluralism (see chapter 1).

And from this observation there is only a single inference required to arrive at the conclusion that we should apply Mayr's lemma to the social behavior of humans: there is no obvious metaphysical difference between the social behavior of our species and the social behavior of many other species, and since Mayr's lemma has held true for all scientific theories that have been generated to explain, inter alia, the social behavior of these nonhuman species, it is probably the case that Mayr's lemma applies to our attempts explain and understand the social behavior of the species that we are as well. Technically, this requires we adjust our formulation of Mayr's lemma to account for the fact that the social sciences do not have a working definition of the construct *ultimate explanation*. That is easiest done by reformulating Mayr's lemma as a slightly more general claim to be applied as a special case to ultimate explanations in biology. The reformulated and moral general principle says that scientifically plausible structural explanations and proximate explanations stand in a many-to-many relationship.

References

Alcock, J. 2001. *The Triumph of Sociobiology*. Oxford: Oxford University Press.

Alcock, John, and Paul Sherman. 1994. "The Utility of the Proximate-Ultimate Dichotomy in Ethology." *Ethology*. Formerly *Zeitschrift Fur Tierpsychologie* 96 (1): 58–62.

Blumstein, Daniel T., and Kenneth B. Armitage. 1997. "Alarm Calling in Yellow-Bellied Marmots: I. The Meaning of Situationally Variable Alarm Calls. " *Animal Behaviour* 53 (1): 143–171.

Blumstein, Daniel T., Jeff Steinmetz, Kenneth B. Armitage, and Janice C. Daniel. 1997. "Alarm Calling in Yellow-Bellied Marmots: II. The Importance of Direct Fitness." *Animal Behaviour* 53 (1): 173–184.

Bouchet, P., and J. P. Rocroi. 2005. *Classification and Nomenclator of Gastropod Families*. West Falmouth, MA: Institute of Malacology.

Boyd, Robert, and Peter J. Richerson. 2009. "Culture and the Evolution of Human Cooperation." *Philosophical Transactions of the Royal Society of London. Series B, Biological Sciences* 364 (1533): 3281–3288.

Cooke, Samantha, Dieta Hanson, Yayoi Hirano, Elysse Ornelas-Gatdula, Terrence M. Gosliner, Alexey V. Chernyshev, and Ángel Valdés. 2014. "Cryptic Diversity of Melanochlamys Sea Slugs (Gastropoda, Aglajidae) in the North Pacific." *Zoologica Scripta* 43 (4): 351–369.

Cushman, Fiery, and Joshua D. Greene. 2012. "Finding Faults: How Moral Dilemmas Illuminate Cognitive Structure." *Social Neuroscience* 7 (3): 269–279.

Daly, M., and M. Wilson. 1988. *Homicide. Evolutionary Foundations of Human Behavior*. Piscataway, NJ: Transaction Publishers.

Dewsbury, Donald A. 1994. "On the Utility of the Proximate-Ultimate Distinction in the Study of Animal Behavior." *Ethology*. Formerly *Zeitschrift Fur Tierpsychologie* 96 (1): 63–68.

Flaxman, S. M., and P. W. Sherman. 2000. "Morning Sickness: A Mechanism for Protecting Mother and Embryo." *Quarterly Review of Biology* 75 (2): 113–148.

Gardner, Andy. 2013. "Ultimate Explanations Concern the Adaptive Rationale for Organism Design." *Biology & Philosophy* 28 (5): 787–791.

Gould, S. J., and R. C. Lewontin. 1979. "The Spandrels of San Marco and the Panglossian Paradigm: A Critique of the Adaptationist Programme." *Proceedings of the Royal Society of London. Series B, Containing Papers of a Biological Character. Royal Society* 205 (1161): 581–598.

Greene, Joshua, and Jonathan Haidt. 2002. "How (and Where) Does Moral Judgment Work?" *Trends in Cognitive Sciences* 6 (12): 517–523.

Haidt, Jonathan. 2001. "The Emotional Dog and Its Rational Tail: A Social Intuitionist Approach to Moral Judgment." *Psychological Review* 108 (4): 814–834.

Haidt, Jonathan. 2007. "The New Synthesis in Moral Psychology." *Science* 316 (5827): 998–1002.

Hauser, Marc, Fiery Cushman, Liane Young, R. Kang-Xing Jin, and John Mikhail. 2007. "A Dissociation Between Moral Judgments and Justifications." *Mind & Language* 22 (1): 1–21.

Hauser, Marc D. 2006. "The Liver and the Moral Organ." *Social Cognitive and Affective Neuroscience* 1 (3): 214–220.

Jablonka, E., M. J. Lamb, and A. Zeligowski. 2014. *Evolution in Four Dimensions, Revised Edition: Genetic, Epigenetic, Behavioral, and Symbolic Variation in the History of Life*. Cambridge MA: MIT Press.

Jones, Frank Morton. 1932. "Insect Coloration and the Relative Acceptability of Insects to Birds." *Transactions of the Royal Entomological Society of London* 80 (2): 345–771.

Joyce, R. 2007. *The Evolution of Morality*. Cambridge, MA: MIT Press.

Krajíček, Jan, Petr Kozlík, Alice Exnerová, Pavel Stys, Miroslava Bursová, Radomír Cabala, and Zuzana Bosáková. 2014. "Capillary Electrophoresis of Pterin Derivatives Responsible for the Warning Coloration of Heteroptera." *Journal of Chromatography A*: 94–100.

Krebs, Dennis L. 2008. "Morality: An Evolutionary Account." *Perspectives on Psychological Science: A Journal of the Association for Psychological Science* 3 (3): 149–172.

Machery, Edouard. Forthcoming. "Discovery and Confirmation in Evolutionary Psychology." In *The Oxford Handbook of Philosophy of Psychology*, ed. Jesse Prinz.

Martín, José, and Pilar López. 2006. "Vitamin D Supplementation Increases the Attractiveness of Males' Scent for Female Iberian Rock Lizards." *Philosophical Transactions of the Royal Society of London. Series B, Biological Sciences* 273 (1601): 2619–2624.

Mayr, Ernst. 1961. "Cause and Effect in Biology: Kinds of Causes, Predictability, and Teleology Are Viewed by a Practicing Biologist." *Science* 134 (3489): 1501–1506.

McKenzie Alexander, J. 2010. *The Structural Evolution of Morality*. Reissue edition. London: Cambridge University Press.

Nei, Masatoshi, Yoshiyuki Suzuki, and Masafumi Nozawa. 2010. "The Neutral Theory of Molecular Evolution in the Genomic Era." *Annual Review of Genomics and Human Genetics* 11 (1): 265–289.

Nowak, Martin A. 2006. "Five Rules for the Evolution of Cooperation." *Science* 314 (5805): 1560–1563.

Ornelas-Gatdula, Elysse, Anne Dupont, and Ángel Valdés. 2011. "The Tail Tells the Tale: Taxonomy and Biogeography of Some Atlantic Chelidonura (Gastropoda: Cephalaspidea: Aglajidae) Inferred from Nuclear and Mitochondrial Gene Data." *Zoological Journal of the Linnean Society* 163 (4): 1077–1095.

Scott-Phillips, Thomas C., Thomas E. Dickins, and Stuart A. West. 2011. "Evolutionary Theory and the Ultimate–Proximate Distinction in the Human Behavioral Sciences." *Perspectives on Psychological Science: A Journal of the Association for Psychological Science* 6 (1): 38–47.

Sherman, Paul W. 1985. "Alarm Calls of Belding's Ground Squirrels to Aerial Predators: Nepotism or Self-Preservation?" *Behavioral Ecology and Sociobiology* 17 (4): 313–323.

Shriner, Walter M. 1998. "Yellow-Bellied Marmot and Golden-Mantled Ground Squirrel Responses to Heterospecific Alarm Calls." *Animal Behaviour* 55 (3): 529–536.

Smith, Maynard J., and D. G. C. Harper. 1995. "Animal Signals: Models and Terminology." *Journal of Theoretical Biology* 177 (3): 305–311.

Thompson, T. E. 1960. "Defensive Acid-Secretion in Marine Gastropods." *Journal of the Marine Biological Association of the United Kingdom* 39 (1): 115–122.

Thornhill, Randy, Steven W. Gangestad, Robert Miller, Glenn Scheyd, Julie K. McCollough, and Melissa Franklin. 2003. "Major Histocompatibility Complex Genes, Symmetry, and Body Scent Attractiveness in Men and Women." *Behavioral Ecology* 14 (5). ISBE: 668–678.

Tinbergen, N. 1963. "On Aims and Methods of Ethology." *Zeitschrift Für Tierpsychologie* 20 (4): 410–433.

West-Eberhard, Mary Jane. 2003. *Developmental Plasticity and Evolution*. Oxford: Oxford University Press.

Wilson, David Sloan. 2004. "What Is Wrong with Absolute Individual Fitness?" *Trends in Ecology & Evolution* 19 (5): 245–248.

Wilson, E. O. 2000. *Sociobiology: The New Synthesis*. Cambridge, MA: Belknap Press of Harvard University Press.

3 The Social and the Psychological Dimensions of Ethics

Our focus now switches to the relationship between philosophical ethics and moral psychology. For the latter, of course, I am referring to experimental—or at least predominantly empirical—research undertaken by people with graduate training in the behavioral or social sciences, and which focuses on some psychological dimension of moral cognition, choice, or behavior. I am not proposing to analyze, in other words, the relationship between philosophical ethics and philosophical moral psychology.

How shall we proceed? I will begin by introducing a conceptual analysis of sports games—games, that is, with official, explicit rules, and established public protocols for creating, evaluating, or otherwise modifying those rules—and then show that the components of this analysis can also be used to pick out roughly corresponding distinctions in several of the most plausible theories in philosophical ethics. The strategy here is straightforward. The analytical concepts that are able to refer to some of the fundamental aspects of either possible or actual sports games can also be used, more generally, to talk about the fundamentals of other forms of norm-guided or norm-regulated, or even norm-realizing, social behavior, and where, importantly, some of the relevant behaviors are well within the scope of philosophical ethics. But I should stress here that my aim is not scholarly: I am not trying to introduce a new way of reading Mill and then relating this reading to another interpretation of Kant. Rather, the point of the analysis is to introduce some novel conceptual resources that can be used to explicitly describe the relationship between philosophical ethics and moral psychology. And in a nutshell, I propose that philosophical ethics is about the explicit norms of possible social behaviors that are metaphysically analogous to sports games in certain respects, while moral psychology is about the proximate processes that are metaphysically analogous to the psychological and development processes, reference to which explains how sports are learned and which also explains what goes on in the bodies and heads of the players when the relevant game is played.

Why sports? Games of this type have many attractive properties for those of us who are interested in understanding the relationship between norms and real social behavior. Unlike many other kinds of norm-regulated behavior, for sports we already know what the norms are: they are the explicit, published rules. Games that are not sports are, arguably, games that have implicit or merely conventional rules—since there are rarely explicit protocols that can be followed to modify or change the rules, it is less than obvious that the rules of nonsports games have public standards of justification. Sports are also much simpler than some other examples where we also know at least some of the official rules and have a good sense of how these rules are evaluated and justified, such as certain areas of law, medical ethics, and perhaps even the norms of data analysis in certain areas of scientific research. What's more, there are no deep metaphysical problems presented about the existence of sport—games of this kind are well within the ontology of perhaps every single living human alive; if you think about it, that is a pretty rare thing. And finally, please note that—recalling from the introduction the definition of realizable norms—playing a sport ipso facto is to *realize* the norms of the game.

So, hereafter allow me to just say "games" when I should say "games that are sports." There is one further, important constraint to take on board before we proceed into the details of the relevant analysis. It is important that our conceptual understanding of games be compatible with Mayr's lemma—that our analysis not, for instance, entail that it will routinely be possible to deduce truths about the structure of the game from truths about the physiological or psychological (proximate) mechanisms involved in playing the relevant game. Of course, the playing of almost any game is a paradigmatic example of the complex interaction between structural and proximate factors; the argument is not that, in reality, structural and proximate processes are causally separated from one another. Instead, Mayr's lemma holds at the level of the epistemology of inquiry: it applies only to the hypotheses and explanation we formulate in order to understand a complex social behavior like the playing of hockey or football or soccer. Mayr's lemma tells us why our initial descriptions of the rules, strategies, game positions, and so on will not yield any particularly deep psychological insights into how people learn and play the relevant game, and why our initial best guesses about how the rules of the game are represented psychologically will not provide deep insight into what the optimal strategy is for playing and winning the relevant game, or even if there is an optimal strategy at all.

The Social and Psychological Parts of Games

Wittgenstein writes in the introduction to *Philosophical Investigations*, "One can also imagine someone's having learnt the game without ever learning or formulating rules. He might have learnt quite simple board-games first, by watching, and have progressed to more and more complicated ones" (Wittgenstein, Hacker, and Schulte 2010, lix). Wittgenstein is right. It is possible to know how to play games without also knowing the exact formal or explicit rules for the game. That is, it is possible—and for some games quite routine—for the public content of the official rules to be very different from the private psychological content of the beliefs, attitudes, and other mental representations that constitute a player's knowledge of the relevant game. This difference between the public rules and the psychological content does not impair the player's ability to play the game, and may for some games be the reason why players excel at them, such as when the official rules are very complicated or very numerous, and players instead rely on heuristics or some other psychological shorthand.

It obviously is a very difficult empirical question to determine the structure and content of psychological representations motivating any reasonably complex social behavior. But in order to arrive at the first pair of concepts in our analysis, all that matters is that we make a distinction between the content of the public or official rules for a game and the content of the psychological representations that players have in virtue of which they are able to play the game and thereby realize the rules. What terms can capture this distinction? First, we have the concept of a *regulative framework*. The official rules of a sport like hockey or Formula 1 racing or Nine Men's Morris are a set of rules that form a regulative framework. I say "framework" because no one rule—assuming that rules are individuated by the sentences used to express them, usually of the form "S should φ when condition C is true"—is able to give rise to any of these games alone. Moreover, collecting all of the particular rules of a game together into some super rule—by, for instance, forming a very large conjunction out of each rule—would break the rules' ability to describe and thus regulate the game. Why? The rules of a regulative framework apply with varying degrees of strength and, more importantly, under logically independent conditions that can obtain throughout the course of a game. In some states of the game, certain rules are applicable while others are inapplicable, but then, in other states, the inapplicable rules become applicable and the initially applicable rules lose their force. However, it will not be possible to anticipate a priori all of the possible states of a game, so it will not be possible to state the rules of the game as a (very) complex sentence. Rather, it is the dynamic structure and the history of the particular game as it is being played that partly determine which of the rules are to be followed at a particular time. The rules must be formulated in ways that are sensitive to the states and transformations that the game both can and cannot enter and proceed through, and the players must be intelligent enough to figure out how to fit the rules to the present state of the game. So, regulative frameworks have conditions of application that prevent interpreting the rules of a game as a single sentence, for no rule that applies all at once will be able to act as a regulative framework for a social action with any more than a single state. Because of that, it makes sense to say that regulative frameworks are constituted by clusters of rules or norms with differing conditions of application.

But what is the difference between a rule and a norm? Following Jon Elster (Elster 1989), a simple way of distinguishing between rules and norms is to say that norms are a special kind of rule: norms are agent-neutral rules. Rules can be for particular people ("Johnny goes to bed at 9:00 p.m."), but norms cannot refer to any particular person, though they can refer to the social roles they occupy ("Children go to bed at 9:00 p.m."). To define regulative frameworks as constituted by clusters of norms is therefore to define regulative frameworks as inherently public entities, in a way that systems of rules are sometimes not. Someone who is trying on their own to lose weight may make for themselves a list of rules to follow; but according to the definitions we are developing here, these rules do not technically form a regulative framework because these rules are not norms. However, someone trying to design a public weight-loss program for their community will, technically, be thinking about how to build a regulative framework out of clusters of norms. And of course, games of sport are meant to be public, because, by design, the rules that form the regulative framework of any sport have not been formulated so that only specific people can realize them.

Now, an individual norm may be part of several different regulative frameworks, but in order for norms in a regulative framework to describe the actions necessary to enact any reasonably complex pattern of social behavior, these norms must "work together." In other words, and extending our earlier reasoning, no regulative framework can be constituted by a single rule or single norm if it is also to be able to give structure to any form of nontrivial social behavior. The argument for this is that for a regulative framework to describe a process of anything but the most trivial of social actions, it must be sensitive to the states and transformations that component actions proceed through, and different norms will be required to both refer and provide guidance relative to the various possible transformations of the relevant action. There is, however, a less abstract way to introduce the same underlying philosophical observation. Imagine trying to create an entirely new game that is defined by only a single norm. If you tried to teach someone this new game, they would almost immediately begin to ask a series of "what if" questions, and answers to these questions would amount to the formulation of new norms for the game. Practically speaking, then, it takes networks or clusters of interrelated norms to create regulative frameworks that are useful for defining social behaviors.

Because some norms are impossible to realize, by the definitions we have introduced so far, it must also be true that some regulative frameworks are impossible to realize. It could be, for instance, that the social behaviors described by the relevant regulative framework are too complicated for people to enact, or the relevant behaviors may only be possible if at least some people were twice as strong or smart or kind as even the smartest, strongest, or kindest people currently alive. Put another way, designers of a regulative framework are not logically bound to propose only frameworks of norms that are, technically, realizable—and one of the metaphysical reasons why a particular framework may not be realizable is that it would only be realizable if a set of false contingent facts about human abilities, motivations, talents, and so on were actually true. We can call these facts the *physical presuppositions* of a regulative framework. These are the properties that people must have in order for it to be possible to realize the regulative framework in question. Every regulative framework will have a set of physical presuppositions, but it does not follow that the set of physical presuppositions for each regulative framework is true, because one (and not the only possible) reason why a regulative framework cannot be realized is that most of its physical presuppositions are false.

The concept of a physical presupposition is nicely illustrated by the relationship between the ability to skate and the regulative framework of hockey. The official norms of hockey do not say that humans can skate, and neither are the norms of hockey anything like a physiological theory that explains how humans skate. The rule book for professional hockey provides no instruction at all on how to skate. Nevertheless, in order for the regulative framework of hockey to be put into action, it must be the case that players of hockey are able to skate—and survive being checked, and be able to track the relative positions of both teammates and opponents and a little black disc, and so on. Thus, there is a counterfactual dependence between the realization of a regulative framework (playing a game of hockey) and a set of abilities (the physical presuppositions of the game, such as the ability to skate). If the physical presuppositions of a regulative framework are false, then it is impossible to realize the relevant framework. Consequently, we can investigate whether or not it is practically possible to realize a particular regulative framework by investigating both what its physical presuppositions happen to be, and also whether or not a sufficient number of these presuppositions are true.

We have so far the concepts of a regulative framework and its physical presuppositions; we began with Wittgenstein's observation that propositional content of the rules of a game—or, for our purposes, the norms of a regulative framework—can be different from the propositional content of the knowledge in virtue of which people are able to play the relevant game. To capture this distinction, we can introduce the concept of the *psychological representation* of a regulative framework. This refers to the knowledge people have that allows them to realize the relevant regulative framework. Since there is a counterfactual dependence between the existence of the right psychological representation and the realization of the corresponding regulative framework, the psychological representation of a regulative framework is similar the framework's physical presuppositions. However, the key difference is that any psychological representation is inherently *mentalistic*: this concept refers only to the beliefs, desires, theories, attitudes, and whatever other mental or cognitive entities are necessary in order for someone to know how to realize a particular social practice.

So what should we call the rules of the game—the norms of the framework—as they are explicitly represented in public representations of the regulative framework? We can say that the norms are the *structural content* of the framework. When expressed linguistically, the structural content of a regulative framework can be thought of as third-person descriptions of patterns of social behavior. For games that are not realized, the structural content of the framework would be expressed as an inferential network of sentences that mostly have "should" (or "most" or any similar deontic term) as the lexical verb. When the relevant game is realized, then while the structural content of the regulative framework will be unchanged, an additional set of sentences that mostly have the complex construction "both should and it is the case that" will be true, and where what follows this construction is the same as what follows the "should" in the linguistic expression of the structural content of the regulative framework.

I hope you can now see that we have a set of analytical concepts that mirrors how the distinction between structural and proximate explanations operates in biological theory. Regulative frameworks can be defined and identified in terms of their structural content, psychological content, and physical presuppositions. At the same time, knowing the details of, say, the structural content of some particular regulative framework will not provide any a priori insights into either the psychological representation or physical presuppositions of the relevant framework. Put another way, all three aspects of regulative frameworks must be investigated using three different conceptual vocabularies. For hockey, there would be the respective vocabularies of cognitive psychology (psychological representation), the normative constructs proprietary to hockey, like "blue line" and "icing" (the structural content), and physiology (physical presuppositions). Furthermore, and more deeply, because confirming whatever psychological hypotheses accurately represent what professional hockey players know in virtue of which they are able to play games of hockey will provide little insight into which specific physiological hypotheses about the very same hockey players are also true, scientifically plausible hypotheses about the psychological representations and physical presuppositions that hold of real games of hockey likely stand in a many-to-many relationship.

There is one remaining concept to introduce. This concept is the *enabling knowledge* of a regulative framework. Enabling knowledge is not part of the psychological representation of the norms of a regulative framework, and neither is a component of the different skills or abilities that group of people must have if a regulative framework can be realized by their behavior. It is, instead, whatever practical knowledge (or phronēsis) a person has in virtue of which they are able to "do better than literal competence" when a framework is put into action. In other words, enabling knowledge is what some people have in virtue of which they can perform some kind of social action with proficiency, or virtuosity, or talent, or mastery—in short: *excellence*. A team of professional hockey players may go a whole season's worth of games and not win even a single time, and yet nevertheless succeed in playing a season of "legally valid" games. These are games in which the norms of hockey are realized in such a way that, for example, the opposing teams never abandon the game mid-way because they realize that the other team simply does not know the rules. The constantly losing team in this example is, despite their losses, still competent at hockey bcause they both know how to play the game and are able to put this knowledge into action. However, when either a team or at least some of its players acquire the additional knowledge that allows them to excel at the game, then we can call the knowledge that allows for this excellence *enabling* knowledge. And so, to put the idea even more concretely, the parents who are teaching their children how to play hockey are mostly interested in whether or not their kids have the right psychological representation of the regulative framework of hockey, while coaches for professional hockey teams are hired partly because of their ability to develop enabling knowledge in their players.

Here now is the full distinction. Regulative frameworks are clusters of norms that both describe and differentiate kinds of complex social actions. Each regulative framework can be identified by its structural content (the norms themselves), psychological representations of its structural content (what people must know in order to enact the regulative framework), its physical presuppositions (the capabilities that people must have in order to put the framework into action), and, sometimes, its enabling knowledge (the additional know-how that people acquire from practicing putting the regulative framework into action that, in turn, can increase or decrease their respective abilities to achieve some of the ends of the regulative framework). My view is that this package of analytical concepts describes some of the fundamental metaphysical dimensions of most sports; without too much further conceptual work, the same analytical framework can also describe one attractive way of relating philosophical ethics to moral psychology.

Applying the Framework to Philosophical Ethics

The first step in establishing this connection is to consider evidence that the concepts of a regulative framework, its structural content, its psychological representation, and its physical presuppositions can be applied, without too much conceptual tension, to some of the most plausible, or at least most widely discussed, theories in the history of philosophical ethics.

And there probably is no theory in Western philosophical ethics more famous than utilitarianism, which makes J. S. Mill's version of utilitarianism a natural place to start. For Mill, the structural content of utilitarianism—which, please keep in mind, are the norms of the theory—are duties that promote happiness. Famously, Mill writes, "actions are right in proportion as they tend to promote happiness, wrong as they tend to produce the reverse of happiness" (Mill 1925, 8). This claim can be read as a formula: we discover which norms are ethical duties by discovering which acts promote happiness. However, Mill also urges his readers not to confuse this list of duties with either (what we are calling) a psychological representation of the duties, or the physical presuppositions of his utilitarianism. Requiring people to act explicitly from knowledge of the rules of utilitarianism "is to mistake the very meaning of a standard of morals and confound the rule of action with the motive of it. It is the business of ethics to tell us what are our duties, or by what test we may know them; but no system of ethics requires that the sole motive of all we do shall be a feeling of duty; on the contrary, ninety-nine hundredths of all our actions are done from other motives, and rightly so done if the rule of duty does not condemn them" (Mill 1925, 17–18).

Therefore, there is in Mill grounds for drawing a distinction between the structural content of utilitarianism and its corresponding psychological representation; for Mill, it is possible (and possibly quite routine) that these have very different propositional contents. Moreover, it is only a single inference from this to observe that people can be better or worse at following utilitarian duties. Perhaps they are comparatively better or worse at attending to which actions are honest and which are graceful, and, through thinking about these qualitative differences, learning about which actions can be appropriate to follow in different situations. That would be some of the enabling knowledge of utilitarianism. Finally, the cognitive ability to understand moral duties is one of the more obvious physical presuppositions of Mill's theory, and the ability to experience happiness is an even more obvious further example. More subtly, we may also want to include the ability to develop positive attitudes toward moral knowledge and motivation among the physical presuppositions of Mill's theory (Hurka 1998). That said, far more substantial physical presuppositions of Mill's own flavor of utilitarianism are what he elsewhere calls universal "social feelings, the desire to be in unity with our fellow creatures" that are "powerful principles" of human nature (Mill 1925, 30).

So, we can find textual support for reading into Mill our distinction between the structural content of a regulative framework, and both its psychological representation and its physical presuppositions, and perhaps even its enabling knowledge. After Mill, it is natural to next consider Kant. The structural content of Kant's ethics is, of course, exactly those rules ("maxims") that can be deduced from applications of the categorical imperative to particular contexts of choice or action. And quite famously his ethics has, among other physical presuppositions, very deep commitments about the autonomy of a rational will. But of particular interest given our aims is the fact that the first formulation of the categorical imperative—"act only in accordance with that maxim through which you can at the same time will that it become a universal law" (Kant [1785] 1998, 31)—appears to describe a decision procedure. Robert Johnson summarizes the procedure this way:

> First, formulate a maxim that enshrines your reason for acting as you propose. Second, recast that maxim as a universal law of nature governing all rational agents, and so as holding that all must, by natural law, act as you yourself propose to act in these circumstances. Third, consider whether your maxim is even conceivable in a world governed by this law of nature. If it is, then, fourth, ask yourself whether you would, or could, rationally *will* to act on your maxim in such a world. If you could, then your action is morally permissible. (Johnson 2004)

This decision procedure—or applications of it in particular circumstances—is obviously central to the psychology of Kant's ethics, since the procedure reflects what it is for a person to come to know his moral duty. Indeed, Kant gives some indication that the reason he provides three different formulations of the categorical imperative is to make it easier to appreciate how to use the imperative as the basis for figuring out one's duty—how to, in other words, produce in one's mind a *psychological representation* of the appropriate norm to follow in a particular context. Following the decision procedure thus is intended to align psychological representations with structural content. Kant writes, "The three ways mentioned of representing the principle of morality are, however, fundamentally only so many formulas of precisely the same law, one of which unites the other two in itself. Nonetheless, there is a variety among them, which is to be sure more subjectively than objectively practical, namely that of bringing an idea of reason nearer to intuition ... and, through this, nearer to feeling" (Kant [1785] 1998, 43).

There is some textual evidence that suggests we can also apply our construction of *enabling knowledge* to Kant's moral philosophy. Kant includes the virtues in his ethical theory, and for him they are the different ways of being cognitively committed to fulfilling one's moral and rational duty in the face of choices or temptations to do otherwise. He writes, "Virtue is *moral strength* in adhering to one's duty, which never should emerge from habit but should always emerge entirely new and original from one's way of thinking" (Kant [1798] 2006, 28). Differences in moral competence are, for Kant, sometimes explained by differences in ability to commit to and act upon one's duty, which is of course one form of the kind of know-how we are calling *enabling knowledge*.

The distinction between the structural content of a regulative framework and psychological representations of this content is also easy to find in the work of Rawls. Indeed, it arguably is a central commitment of Rawls's ethical methodology that decisions about which norms are ethical norms must be made, for them to be fair, by people who are somehow rendered able to reason about their society from a structural perspective *only*. In one of the first papers to introduce his two principles (Rawls 1958), Rawls asks us to imagine that a group composed entirely of rational, self-interested, and nonenvious people gathers to bargain over the fundamental norms of the society that they will subsequently form:

> Their procedure for this [bargaining] is to let each person propose the principles upon which he wishes his complaints to be tried with the understanding that, if acknowledged, the complaints of other will be similarly tried, and that no complaints will be heard at all until everyone is roughly of one mind as to how complaints are to be judged. They each understand further that the principles proposed

and acknowledged on this occasion are binding on future occasions. Thus each will be wary of proposing a principle which would give him a peculiar advantage, in his present circumstances, supposing it to be accepted. (Rawls 1958)

The philosophical insight that follows from these remarks is that a self-interested person would not propose norms favorable only his or her particular interests, because such a norm would be unacceptable to all other people involved in the negotiation who have different interests. Instead, a rational person desiring agreement (and not, that is to say, a failure of the whole deliberative process leading to a failure to form the new society) would be forced to propose only principles that would be acceptable to everyone else involved in the process. Robert Paul Wolff argues that what Rawls was really aiming at here was a game-theoretic proof that people so situated and following such a procedure would arrive at his two principles (Wolff 1977). But of course, Rawls never produced such a proof—and instead he later introduces the device of a veil of ignorance (Rawls 1971), which adds to the bargaining situation the further stipulation that the people bargaining over the fundamental norms of their society must be deprived of any knowledge of their own identity and interests, and are compelled to think only about the fundamental normative *structure* of their society, not their place in it; for our purposes, it does not matter whether or not Rawls is correct to conclude that his two principles would be selected as a result of such a bargaining process.

What principles would be selected by people enjoined to reason about the structural content of their society's most basic regulative framework? Rawls's two principles are as follows:

First Principle: Each person has the same indefeasible claim to a fully adequate scheme of equal basic liberties, which scheme is compatible with the same scheme of liberties for all.

Second Principle: Social and economic inequalities are to satisfy two conditions: they are to be attached to offices and positions open to all under conditions of fair equality of opportunity, and they are to be to the greatest benefit of the least-advantaged members of society. (Rawls 1971, 53)

Like Kant's categorical imperative, these principles can function like a schema for generating more specific principles, and for that reason, it may be reasonable to extend our conception of the structural content of Rawls's regulative framework to include the norms that would be generated by applications of his principles.

So much, then, for the structural content of Rawls's theory; what about the physical presuppositions of Rawls's theory of justice? Famously, Rawls distinguishes between *ideal* and *nonideal* political theory. Ideal political theory seeks to describe goals or ends for a society's institutions, while "nonideal theory asks how this long-term goal might be achieved, or worked toward, usually in gradual steps. It looks for courses of action that are morally permissible and politically possible as well as likely to be effective. (...) So conceived, nonideal theory presupposes that ideal theory is already on hand. For until the ideal is identified, at least in outline—and that is all we should expect—nonideal theory lacks an objective, an aim, by reference to which its queries can be answered" (Rawls 2001, 89–90). What's more, according to Rawls, ideal theory is formulated using an assumption of "strict compliance," and what this means is that, for example, it is simply assumed the regulative framework constructed on the basis of the two principles can be realized both unproblematically and universally—and so, in the conceptual framework we are developing, a theorist is doing ideal theory if (but not only if) she is unconcerned about whether the physical presuppositions of the regulative framework she is analyzing are true. But it does not follow that ideal normative theory lacks physical presuppositions—and indeed, finding out that an ideal theory's physical presuppositions are false would explain why an ideal ethical theory is unrealizable.

It is also worth observing that Rawls may also make a particularly strong physical presupposition in his argument for the two principles. The switch from a bargaining situation in which those present are aware of their own interests to a situation in which the bargainers are deprived of knowledge of their own interests probably reflects a deeper commitment on the part of Rawls to the view that people are fundamentally risk averse (Harsanyi 1975). However, we do not need to answer that question here, because Rawls's writings are liberally salted with various different descriptions of what we can regard as further, though less central, physical presuppositions of his regulative framework. For example:

> This conclusion [that the two principles of justice would be adopted by people in the original position] is strengthened when one adds that the parties regard themselves as having a higher-order interest in how their other interests, even fundamental ones, are regulated and shaped by social institutions. They think of themselves as beings who can choose and revise their final ends and who must preserve their liberty in these matters. A free person is not only one who has final ends which he is free to

pursue or to reject, but also one whose original allegiance and continued devotion to these ends are formed under conditions that are free. (Rawls 1974, 143)

So, again, the question is not whether Rawls is correct in this passage; what is important is only that the construct *physical presupposition* can be applied to Rawls's project, and at minimum the construct picks out whatever psychological or physiological facts still apply to persons behind the veil of ignorance.

Rawls's distinction between ideal and nonideal theory gives us a way of forecasting the psychological representations of the structural content of his ethics. Behind the veil of ignorance, the psychological representation of Rawls's regulative framework will be belief in, or about, the two principles of justice. But this is of course a fictional scenario: there is no sense to be made of the idea that moral psychologists could somehow investigate the minds of people in the original position. More realistically, the psychological representation of a Rawlsian regulative framework would include belief in the two principles plus some complex set of further ideas, motivations, and attitudes that form the cognitive ability to participate in the various different social activities that are, in turn, required to move a society closer and closer toward realizing the two principles. Belief in the two principles would also be part of the theory's psychological representation, but the other beliefs would change as society moves closer toward, or further away, from the ideal.

Our last example is W. D. Ross's ethical intuitionism, and I introduce his ethics as an example of a view that does not fit neatly into the conceptual constructs defined in this chapter. In a well-known passage from his *Foundations of Ethics*, Ross writes, "Both in mathematics and in ethics we have certain crystal-clear intuitions from which we build up all that we know about the nature of numbers and the nature of duty. And ... we do not read off our knowledge of particular branches of duty from a single ideal of the good life, but build up our ideal of the good life from intuitions into particular branches of duty" (Ross [1939] 2000, 144–145). According to Ross, moral intuitions (which he also calls "sincere convictions") of rational and knowledgeable people are the primary data upon which ethical theory is built, and we construct a moral framework for ourselves by using our various moral intuitions as our primary data. Ross differs from Rawls, Kant, and Mill in that, for Ross, the normative content of his ethics is derived almost directly from both personal psychological experience and personal reflection on that experience—so it would be a mistake to say that the norms of Ross's ethics are *structural*. And given that for Ross ethics is mostly a matter of personal intellectual development (the word "social" appears only once in *Foundations of Ethics*), there is no good reason to apply the concepts of physical presuppositions (the capacity to have moral intuitions?) and psychological representation of the duties (the intuitions themselves, plus beliefs derived from reflecting on the intuitions?) to his philosophy. Or, to put the conclusion more generally, the analytical framework developed previously is not well suited to ethical theories that entail that a form of methodological individualism is true of human ethical behavior.

But Ross's individualistic approach to ethics needs to be clearly separated from projects in philosophical ethics that, because they posit no ethical norms and thus no regulative frameworks, cannot be analyzed in our framework—and yet, *precisely because these projects posit no ethical norms*, are nevertheless compatible with the idea that a *different* philosophical project may generate knowledge of ethical norms. That's a very abstract idea, so let me provide an example of what I mean. Speaking very generally, virtue ethics is not about ethical norms, duties, principles, or, indeed, the normative structure of different communities. Instead, its focus is on the developmental processes and habits of mind that lead to happiness, flourishing, or some other personably valuable outcome. Yet, as Mark LeBar has argued, "virtuous agents should and will have reasons to respect deontic constraints" (LeBar 2009, 643), and it may be that a conceptually separate ethical theory rather than a theory in virtues ethics is able to frame a number of well-justified ethical norms. The key to seeing how these theories are compatible is to understand that the virtue theoretic and the deontic theories apply to different aspects of human normative behavior. It is possible, thus, to see concepts developed by philosophical work in the virtue ethics tradition as describing one dimension of human normative behavior (specifically, some of the most morally important developmental and characterological properties), and the concepts of some ethical theory that posits a regulative framework as describing a different dimension of human normative behavior (specifically, what the normative structure of certain groups or communities either is or should be).

However, because of both the important place that Ross's intuitionism occupies in the history of philosophical ethics, and because virtue ethicists typically do not posit ethical norms, we need to refine our definition of "philosophical ethics" slightly. As I will use the term, it should not be taken to refer to *all* projects in philosophical ethics, but rather, only those projects that posit both ethical norms and regulative frameworks.

Summary and Next Steps

We have now a conceptual framework that regiments a distinction between structural and individual (or proximate) processes as dimensions of the social behavior of humans, as well as evidence that the conceptual components of this framework apply well enough to some of the most paradigmatic theories in the history of philosophical ethics.

To summarize, then, these are the concepts that have been introduced so far:

1. Norms are inherently public rules.
2. Regulative frameworks are defined by their *structural content*, which is a cluster of norms.
3. All regulative frameworks, whether they are realized or not, have both *physical presuppositions* and *psychological representations*.
4. The *physical presuppositions* of a regulative framework are those capacities, abilities, and traits that a population of humans must have in order to realize the regulative framework.
5. The *psychological representation* of a regulative framework is the knowledge that must be grasped in order to realize the relevant regulative framework.
6. When regulative frameworks are realized by groups of people, there may also be different kinds of *enabling knowledge* that some or other of these people have, and these explain differences in the competence with which they realize the relevant norms.

One important takeaway is that a description of the structural content of a regulative framework is not ipso facto a description of the psychological content associated with the regulative framework. Sometimes these two constructs may overlap, as we saw was probably the case for beliefs about Rawls's two principles. But a deeper reason for insisting on a distinction between the structural content of a regulative framework and a corresponding psychological representation is that the regulative framework may simply be unrealized. In that case there are no truths about the content or properties of the psychological representation associated with the regulative framework—but since from the perspective of psychology, descriptions of the structural content of regulative frameworks are an abstraction, ensuring that structural content and psychological representations remain logically distinct categories explains, inter alia, how it is possible to investigate or theorize the former without also doing psychology. That is a desirable methodological implication since, of course, most theories in philosophical ethics have, at best, a speculative connection with any aspect of psychological reality.

But going in the other direction, it is also desirable to have a framework for thinking about the relationship between norms and social behavior that is compatible with the fact that we cannot always deduce what norms a group of people are following from even the most detailed study of their thoughts, beliefs, and other motivations. Wittgenstein is right: someone can learn a game, or pattern of fundamentally obligatory social behaviors, without memorizing the norms that describe either the game or the relevant patterns of behavior. And so while there will have to be some kind of causal relationship between the content of a psychological representation of a regulative framework and the framework's structural content, it is also important for these constructs to be kept separate. One of the deeper reasons for this is that the structural content of different kinds of social behaviors is often expressed in concepts and terms that are proprietary to the social world in which the behavior is realized. Think about how easy it is to label new patterns of actions and their (nonpsychological) consequences: "homerun," "icing," "manslaughter," "beneficence," and so on. Then, from the perspective of whichever discipline is responsible for the conceptual vocabulary used to describe the relevant structural content, there can be multiple initially plausible descriptions of the associated psychological representations—and facts about how the regulative framework is realized will only constrain the additional psychological research needed to learn about the details of the underlying psychological representations. The structural facts will not tell a psychologist which specific psychological hypotheses are true.

That is an important methodological implication, because we now have an explicit way of relating philosophical ethics to moral psychology. Here is the proposal: philosophical ethics is the search for the best-justified structural content for whatever regulative framework is the—or at least is an—*ethical* regulative framework. Moral psychology is, accordingly, the study of the psychological representations and possible enabling knowledge supporting the realization of an *ethical* regulative framework. Defined thus, moral psychology and philosophical ethics are autonomous but interdependent areas of inquiry.

Still, we have one outstanding and very significant problem. Not all norms are ethical norms, and so not all regulative frameworks are ethical frameworks. We need criteria that can determine whether or not a norm is an ethical norm, and so whether or not some possible or actual regulative framework is an *ethical* regulative framework. Historically, answering questions like these has been within the purview of philosophical and religious ethics, although there are recent sociological and anthropological answers as well (Keane 2015). Relatedly, a methodological individualist can provide an answer by stipulating that ethical norms must be defined using only psychological concepts, thereby subordinating philosophical ethics to moral psychology. That is the path recently taken by Joshua Greene who writes, "because I am interesting in exploring the possibility that deontology and consequentialism are psychological kinds, I will put aside their conventional philosophical definitions" (Greene 2007, 360). However, over the following chapters, I will stick to the traditional philosophical path. My guiding assumption will be that one of the theoretical goals of philosophical ethics is to develop a largely nonpsychological theory of ethical norms. At the same time, some innovation is called for, if we are also to use the conceptual framework defined in this chapter as a basis for relating philosophical ethics to moral psychology. For, our approach leads to a different but related methodological problem: moral psychology cannot be about the psychological representations associated with *unrealized* regulative frameworks, because, quite simply, then there would be no populations from which to sample. There simply are no naturally occurring populations of Millians, Kantians, or Rawlsians—or, in other words, the best normative theories in philosophical ethics are, technically speaking, unrealized and are therefore not possible subjects of investigation by modern psychological science. Of course, we will spend more time over the coming chapters dealing with both the first and the second problems—let me say here only that my strategy will be to develop an ethical theory that extends some of the core concepts of philosophical ethics so that they apply to real-life cases of the realization of different regulative frameworks, and which therefore tells us on principled grounds which of these realizations count as *ethical* behaviors.

References

Elster, Jon. 1989. "Social Norms and Economic Theory." *The Journal of Economic Perspectives: A Journal of the American Economic Association* 3 (4): 99–117.

Greene, J. D. 2007. "The Secret Joke of Kant's Soul." In *Moral Psychology: Historical and Contemporary Readings*, ed. T. Nadelhoffer and E. Nahmias, 359–372. New York: Wiley.

Harsanyi, John C. 1975. "Can the Maximin Principle Serve as a Basis for Morality? A Critique of John Rawls's Theory." *The American Political Science Review* 69 (2): 594–606.

Hurka, Thomas. 1998. "Two Kinds of Organic Unity." *The Journal of Ethics* 2 (4): 299–320.

Johnson, Robert. 2004. "Kant's Moral Philosophy." *Stanford Encyclopedia of Philosophy*, February.

Kant, Immanuel. [1785] 1998. *Kant: Groundwork of the Metaphysics of Morals*, ed. Mary J. Gregor. New York: Cambridge University Press.

Kant, Immanuel. [1798] 2006. *Kant: Anthropology from a Pragmatic Point of View*, ed. Robert B. Louden. New York: Cambridge University Press.

Keane, Webb. 2015. *Ethical Life: Its Natural and Social Histories*. Princeton, NJ: Princeton University Press.

LeBar, Mark. 2009. "Virtue Ethics and Deontic Constraints." *Ethics* 119 (4): 642–671.

Mill, J. S. 1925. *Utilitarianism*. Raleigh, NC: Hayes Barton Press.

Rawls, John. 1958. "Justice as Fairness." *The Philosophical Review* 67 (2): 164–194.

Rawls, John. 1971. *A Theory of Justice*. Cambridge, MA: Harvard University Press.

Rawls, John. 1974. "Some Reasons for the Maximin Criterion." *The American Economic Review* 64 (2): 141–146.

Rawls, John. 2001. *The Law of Peoples: With "The Idea of Public Reason Revisited."* Cambridge, MA: Harvard University Press.

Ross, W. David. [1939] 2000. *Foundations of Ethics*. Oxford: Oxford University Press.

Wittgenstein, L., P. M. S. Hacker, and J. Schulte. 2010. *Philosophical Investigations*. New York: Wiley.

Wolff, Robert Paul. 1977. *Understanding Rawls: A Reconstruction and Critique of "A Theory of Justice."* Princeton, NJ: Princeton University Press.

4 Methodological Lessons from Philosophical Ethics

Suppose we accept that certain important projects in the history of philosophical ethics can be fairly interpreted as aiming to identify and justify the structural content of the (or at least an) ethical framework; and suppose that we also accept that moral psychology could then be the investigation of the psychological representation and perhaps also enabling knowledge associated with realizations of any ethical regulative framework. What are some methodological implications of adopting these two positions?

At minimum, this approach requires us to not assume a priori that the conceptual vocabulary useful for determining which norms either are or should be central to any ethical regulative framework will be one and the same conceptual vocabulary as the one that is also useful for describing the details of any psychological representation of the structural content of the relevant framework. That would be like assuming that the vocabulary of natural selection explanations must be the same as the vocabulary of neurophysiology or sociolinguistics. Indeed, induction over other biological and social sciences suggests a stronger conclusion, that ethicists and moral psychologists should accept as one of their default methodological assumptions the view that the concepts of ethics and the language of moral psychology cannot and should not be interdefined. We first encountered the the basis of the argument for exactly that conclusion in chapter 2: the complexity of human social and normative behavior makes it probable that multiple different conceptual vocabularies will be necessary to explain and understand our behavior, just as is the case for scientific theories of the social behavior of other animals. A theory of human ethics and morality that deliberately mirrors the historical development of the biological sciences is unlikely to be, literally, a single conceptually unified theory. Instead, it will most likely be a network of different theories and models and other representational vehicles that apply in different ways to different forms of sociality and organization, and which for these differences will never condense, under even the most intense philosophical pressure, into a single set of logically closed concepts and theorems that can solve, inter alia, the problem of moral skepticism, instruct anyone about what to decide to do when faced with any practical choice whatsoever, prove itself to be compatible with the moral intuitions of professional philosophers, determine the appropriate balance between policies that distribute scarce resources to all members of a society and property rights, and so on. Indeed, perhaps the most we should hope for as a theory of ethics and morality is piecemeal evidence about how to make incremental improvements to different pockets of complexity in across the various social worlds.

But I am starting to digress. The task for this chapter is only to show that there are three methodological principles that, if followed, would help ensure that moral psychology remains consilient with philosophical ethics.

Moral Psychology and Normative Ethics

Using the concepts introduced in the last chapter, and the fact that they apply to some paradigmatic ethical theories, we can say that philosophical ethics is sometimes the study of the structural content of ethical regulative frameworks. Theories in ethics will be associated with psychological theories about the associated psychological representations of the structural content and the physical presuppositions of the framework. But it is also important to add to this the following clarification. Normative theories are not coextensive with the structural content of the regulative frameworks they evaluate. Compared to the output of works in religious ethics (like, say, the 613 rules in the *Sefer Hamitzvot*), philosophers in the Western tradition have not produced that many norms. Instead, as we saw in chapter 3, most philosophical energy has gone toward justifying a small number of fundamental norms, or even just one norm, that can be applied across many different contexts of choice and action. The extremely general scope of application of, for example, Mill's duties or Kant's categorical imperative makes these principles similar to Newtonian laws of nature, and thereby obviates the need for many specific "local" norms. But the important point here is just that the structural content of a regulative framework should not be confused with all of the further inferences, observations, and conjecture that can constitute philosophical arguments for norms in the relevant framework. Philosophical ethics may be about regulative frameworks, but that does not mean that the content of most theories in philosophical ethics are even very indirect descriptions of the structural content of the regulative framework that is under study.

Now, the arguments that a moral philosopher advances for his or her ethical theory will rarely be neutral with respect to both the content of the associated psychological representation and the details of the associated physical presuppositions. It is easy to find examples, such as Mill and Rawls, of moral philosophers who use some psychological premises in sophisticated arguments for their normative conclusions. However, it is not usually the business of a moral philosopher to determine whether these psychological premises are true. After all, most moral philosophers advance arguments for regulative frameworks the norms of which are nowhere realized—and this leads to the question of whether, as a matter of logic only, the psychological premises in a philosophical argument for an ethical theory must be true in order for the relevant theory to also be realizable. The answer to that question is "no," and to see why, imagine the possible world in which people do not naturally care at all about either their own pleasure or that of others, but people do care about paying as little tax as possible, and where the tax code for their society is designed such that habitually applying the principle of utility is the only meaningful way to lower one's personal tax burden. Put more generally, the point is that psychological hedonism does not *necessarily* have to be true in order for Millian utilitarianism to be realized.

That conclusion implies that one class of questions that moral psychology can investigate concerns whether or not the psychological premises employed by different moral philosophers are true. But there is another larger set of questions that includes a subset of purely psychological questions; the larger set is composed of questions asking how it would be possible to realize the regulative framework justified, or at least described, by a theory in philosophical ethics. The nonpsychological questions in this set are, of course, about the social (political, economic, religious) conditions that would also have to obtain in order for the relevant framework to be realized.

Now please consider the following question: Is it possible to design a psychological experiment that is able to differentiate ethical theories according to either psychological representations of their structural content or their physical presuppositions? The answer to this question is yes. Just as the psychology of NHL hockey players will differ subtly from the psychology of KHL hockey players, so too will the psychology of a real-life Millian differ from the psychology of a real-life Sidgwickian. But then, is it possible to design a psychological experiment the results of which would be to differentiate *families* of ethical theories according to either their psychological content or presuppositions? The answer to this question is no. Investigating the physical presuppositions of an ethical theory can be a reliable way of determining whether it is possible to realize the relevant theory, and testing for a psychological representation of its structural content can be a way of determining whether the regulative framework is indeed realized. But since no two ethical theories have exactly the same psychological representations or physical presuppositions, and since a single ethical theory may have multiple psychological representations depending on the particular way that it is realized in different individuals, it will not typically be possible to individuate (or define) families of ethical theory in terms of either a set of psychological representations or a set of physical presuppositions. For example, while Mill's utilitarianism may be linked with a particular set of psychological representations and physical presuppositions, there is neither a set of representations nor a set of presuppositions that is uniquely associated with all—or even just the most famous—utilitarian theories in philosophical ethics.

Let us dig into these issues in more detail. Suppose we have a single normative theory, NT, that posits a specific regulative framework. Let us also assume that the structural content of NT is exceedingly well justified by NT, so that NT might even be the philosophical theory that tells us which regulative framework is really *the* one true objective ethical regulative framework. Under these assumptions, here are some empirical questions that we can ask about NT:

- How is the structural content of NT represented psychologically? Are the same beliefs and habits of thought always present, or are there multiple different psychological representations of NT? When NT is realized, what do people know in virtue of which this realization is possible? And if NT is not realized, is there some pathway from present beliefs, attitudes, and ideologies to instantiating a psychological representation of NT?

- Is the ability to φ a psychological presupposition of NT? Must people be able to φ in order for NT to be realized? Is another physical ability an effective substitute for φ when putting NT into practice?
- Are there some characteristic ways in which people learn the regulative framework (perhaps even only an approximation of it) at the heart of NT?
- Do particular norms in NT have unique physical presuppositions?
- What are some of the characteristic psychological failures or impediments to realizing NT? What learning processes are most effective for producing psychological representations of NT's structural content?
- Are there different pools of enabling knowledge associated with realizations of NT?

The point is that we can learn quite a lot about the psychological and behavioral basis of NT by pursuing any number of different empirical research programs—but only if we also assume that NT is realized at least to a meaningful approximation by some group of people, or there is a process of psychological development that corresponds to a process of cultural change that leads to the realization of NT, or any number of other specific empirical scenarios. The philosophical point here is simply that none of the preceding questions can be answered a priori.

Now suppose that we give ourselves two normative theories, NT_0 and NT_1, and we are interested in comparing the psychological representation and physical presuppositions associated with each theory's regulative framework. So, let us assume that we ask of both NT_0 and NT_1 all of the previous questions and come thereby to learn for both NT_0 and NT_1 what the relevant physical presuppositions and psychological representations are. Because of that work, we do *not* have to do an empirical research study to answer these comparative questions, since the answers can be easily inferred from the existing empirical findings.

- What physical presuppositions, if any, do NT_0 and NT_1 have in common?
- What psychological content, if any, do NT_0 and NT_1 have in common?

The next point, then, is that we can begin to construct a taxonomy of normative theories in terms of their grounding psychologies and physiologies only after a process of meaningful empirical investigation of individual normative theories. We cannot answer either of the two comparative questions, or other, similar questions in the absence of empirical answers to some of the specific questions about individual normative theories listed previously.

But there is also a problem with the assumptions governing the example in which we compared NT_0 and NT_1. To answer the first list of empirical questions, both NT_0 and NT_1 would have to be realized. However, the only ways that it would be possible to realize both NT_0 and NT_1 is if different populations of people realized NT_0 and NT_1 and so form two separate normatively integrated populations; or if the regulative frameworks in NT_0 and NT_1 are compatible and a single population could realize both NT_0 and NT_1. Yet, since different versions of utilitarianism must differ in terms of the structural content of the relevant regulative frameworks, and what it is for a regulative framework to be an *ethical* regulative framework is, plausibly, just for it to have priority over all other realizable regulative frameworks in some context of action, it is a mistake to think that utilitarianism—which is a large family of different normative theories that posit different regulative frameworks—can be realized. Doing so would require that each regulative framework in the family be concurrently realized in one and the same context of action, which is impossible.

Now, one objection might be that it is overreaching to claim that the psychological basis of a normative theory can only be investigated empirically. After all, in chapter 3 did we not arrive at conclusions about the psychological content and presuppositions of different paradigmatic normative theories? We did. But those philosophical conclusions should not be mistaken for an empirical confirmation that, for instance, any of the physical presuppositions we extrapolated from Kant, Mill, or Rawls are true—or that they are necessary for the realization of the relevant normative theories. The argument in chapter 3 showed only that the concepts of a regulative framework and its structural content, psychological representation, and physical presuppositions could be applied without too much conceptual tension to some of the most plausible theories in philosophical ethics. And that argument does not also justify any of the much stronger claims to the effect that, say, the physical presuppositions we explored are in fact true. Maybe people who are not maximally risk averse would agree to Rawls's two principles; maybe the veil of ignorance is completely unnecessary for people to find their way toward an agreement with his two principles. Or maybe there is a way to base decisions upon the categorical imperative without going through something like Kant's decision procedure, or ways for Mill's utilitarian duties to be enacted without also developing empathy. The more important—and more general—methodological lesson to draw from these remarks is that a clear understanding of a normative theory in philosophical ethics helps a moral psychologist to know which specific empirical questions to ask; at the same time, because the psychological representations and physical presuppositions of a regulative framework theory are not logical entailments of the structural content of the relevant theory, these questions cannot be settled by the philosophical arguments that otherwise justify the theory.

Indeed, philosophers often rely upon claims about the physical presuppositions of their theory in advancing arguments for particular norms or regulative frameworks. As we have seen, it has counted for Kant and Mill, as well as Ross and Rawls that they can link their proposed norms with putative facts about human psychology. Yet the validity of their arguments does not mean, of course, that the relevant psychological facts are true. Moral philosophers are mostly, and notoriously, equally speculative and optimistic psychologists, and it may be that Mill and Kant were both reasoning from false and not-even-remotely-close-to-the-truth psychological premises. Nevertheless, even that possibility does not imply that either philosopher's theory cannot be realized, or that the structural content of either theory lacks adequate justification.

Let us reword slightly the second question asked earlier: is it possible to design a psychological experiment the results of which will be able to differentiate *families* of ethical theories according to either their psychological representations or physical presuppositions? The answer, again, is no: families of normative theory cannot be distinguished by unique sets of representations or presuppositions. As an empirical matter, some beliefs will be among the psychological representations of some normative theories and not others, while some cognitive and emotional abilities will be among the physical presuppositions of many different normative theories, even when the relevant theories are both solidly within the same philosophical tradition. Thus, we can ask whether people already implement a particular normative theory, like Mill's utilitarianism or Kant's deontology; but it is a mistake to believe also that *all*—or even more than just one—member of a family of normative theories can be realized at any particular time. Yet because of this particular fact, there cannot be a test that shows, for instance, that people are naturally either utilitarians or deontologists.

This is a difficult line of reasoning for an important conclusion, so let me illustrate it with an example. Suppose that we want to determine the exact details of a group of people's psychological representation of the norms of professional hockey. Given enough time, labor, and money, this is a patently viable psychological project. We have observed people playing hockey, and we have the rule book for the game, and we know when and where and by whom the regulative framework of hockey is realized. We even know what the structural content of the game happens to be, and so we do not have to train people to play hockey in order to begin our research—since there are plenty of naturally occurring games of hockey. However—and this is the key point—we cannot carry out a similar empirical investigation into the psychological representation or physical presuppositions of winter sports, even though millions and millions of people play this family of sports every day. The key difference here is that there is no specific regulative framework associated with the construct *winter sports*; it is instead a construct that names a category of different and not all mutually realizable regulative frameworks. Though groups of people can play games of hockey, curling, or luge, there is no "metasport" that is being played when all individual winter sports are played, nor is it possible for one and the same group of people to realize, concurrently, a game of hockey and several rounds of bobsled and luge racing.

Likewise, examining only the beliefs of a particular group of people involved in some particular social activity will not necessarily yield a structural description of the social activity. That is: it is only because we have a prior and independent structural description of games of hockey that we can say that our studies of the representations and abilities of hockey players are connected with games of hockey. Likewise: it is only because we have a prior structural theory describing a process of natural selection that we can say that some particular developmental process is an adaptation.

The same conclusions apply, mutatis mutandis, to the relation between philosophical ethics and moral psychology. While there are different formulations of utilitarianism and deontology corresponding to different specific regulative frameworks, there is not a "meta-regulative framework" that is realized when some people put into practice either any particular form of utilitarianism, or Kantian ethics, or almost any other normative theory discussed in the history of Western philosophical ethics. "Utilitarianism" is a generic name for a family of different ethical theories that posit distinct regulative frameworks, and the same is true for "deontology." And, just as it makes no sense to inquire into the physical presupposition of winter sports (because there are not any), it makes no sense to inquire into the psychological representations or physical presuppositions of either utilitarianism or deontology. It is simply a category mistake to believe that either utilitarianism or deontology, when understood as two families of ethical theory, has distinctive physical presuppositions or psychological representations.

Two Methodological Lessons

We are now in a position to compile the various insights and implications covered so far into two methodological lessons.

1. *It is a category mistake to try to investigate the psychological representations or physical presuppositions of families of ethical theory. In order to avoid this mistake, researchers in moral psychology should have a reasonably precise understanding of the structural content of the ethical theory that is inspiring their investigation.* This is a sound principle, again, because all members of a family of normative theory will not all have the same unique physical presuppositions or representation of their structural contents. More importantly, this principle remains sound even when we take into account the fact that different normative theories sometimes have overlapping specific presuppositions or psychological representations. Perhaps the most famous example of this is Mill's own sense of how his ethics related to Kant's, for Mill claimed that there is fundamental consistency between how his utilitarian principles and Kant's "first principle of morality" will used by people when they are thinking through everyday ethical actions (Mill 1925, 50–51). And that may be—but it does not follow that Mill's utilitarianism and Kant's ethics are mutually realizable in one and the same social context.
2. *Moral psychology should only investigate the psychological foundations of the most plausible theories in philosophical ethics.* Only the most ideologically simplistic or dogmatically scientistic interpretations of philosophy are able to reach the conclusion that moral philosophy has been, on the whole, a waste of time. For scientists and philosophers not burdened by such attitudes, the efforts of generations of moral philosophers can be understood as producing a reasonably complete picture of the best-justified theories in philosophical ethics. And since it is not possible to investigate either psychological representations or physical presuppositions without also explicitly specifying a guiding normative theory, there is an elegant way of linking philosophical ethics with

> moral psychology: stipulate that moral psychologists can only investigate the psychological foundations of the most plausible theories in philosophical ethics. By holding to that principle, philosophical ethics would support the reliability of moral psychology. Constraining psychological research to the proximate foundations of only the most well-justified theories in philosophical ethics ensures that the selection of ethical theory that gets used to structure research in moral psychology is not ad hoc—or, what would be even worse, is not simply a reflection of the individual morality sensibilities of a principal investigator and her lab that, ironically, gets certified as a plausible ethical theory only because it is a working assumption of scientific research.

This second methodological lesson should be easy to follow in moral psychology. Philosophical research in normative ethics can, to an important degree, often be interpreted as the effort to sort more plausible ethical theories from the less plausible, and this interpretation of moral philosophy does not require believing, furthermore, that philosophical research confirms theories in moral psychology.

Yet, this second methodological lesson is easily violated if the full range of philosophically plausible ethical theories is effectively ignored when particular normative theories are selected for psychological investigation. It is not possible, of course, to investigate the representations and presuppositions of every philosophically plausible ethical theory. But it is important nonetheless to ensure that a representative selection of ethical theories is relied upon as a guide for research in moral psychology. There is, for example, a far greater number of ethical theories than those that are solidly within the classical liberal tradition that are, to say the least, patently philosophically plausible. And even then, there is also a wide range of philosophically plausible interpretations of liberal moral theorists like Kant and Mill (Shanley and Pateman 1991). The underlying methodological idea here is that just as the rigor of a program of experimental science is proportional to the program's ability to ground both its theory and its methodology in the best practices of past research, so too is the rigor of moral psychology proportional to its ability to rely upon philosophical ethics for guidance about which ethical theories are most probably, and perhaps even most likely to be true.

However, the second methodological principle faces a difficult problem at the level of its implementation as a constraint on scientific practice. As we have noted several times already, the most plausible theories in ethical philosophy are nowhere realized. So this principle cannot be implemented by simply sampling from, say, only the avowed Millian utilitarians found, conveniently, among the local population of undergraduates. What's more, it is also one of the ironies of contemporary scientific practice that it would very likely be unethical according to the standards of institutional review boards to try to induce even willing participants to follow, say, Kant's ethics in their day-to-day lives.

With only a few additional assumptions, it is possible to derive the first principle from the second principle. The history of philosophical ethics is both long and deep, and it contains many different ethical theories that are, unsurprisingly, not all equally plausible. There are more and less philosophically credible formulations of utilitarianism and deontology. Consequently, it is not possible to both use estimates of philosophical plausibility to determine which specific ethical theories should guide research in moral psychology and also hold that moral psychology can investigate the psychological foundations of whole families of ethical theories. But unfortunately, some of the methods of the most influential research in contemporary moral psychology reflect that view that it is possible to combine both of these options. As we shall see in chapter 5, it is quite easy to find examples of influential experiments that violate the second methodological principle by seeking to determine whether people really are "utilitarians" or "Kantians," and where this means something quite different from whether or not there are people who are really trying to apply in their lives a well-justified interpretation of either Kant's or Mill's own ethical theory.

Metaethics and Moral Psychology

It is also, however, characteristic of some of the very same research to draw deeply implausible metaethical conclusions. Unfortunately, the relationship between moral psychology and metaethics is not well described by any of the technical concepts that have been introduced so far—so we cannot use these concepts in an argument that explains what exactly is implausible about the relevant metaethical conclusions. That said, the relationship between metaethics and moral psychology is relatively straightforward, and in order for moral psychology to remain compatible with the present state of metaethics, theories in moral psychology must not simply make assumptions about the truth or falsity of metaethical positions that are either independent of, or not controlled for, in the relevant underlying experiments.

In contemporary metaethics, there are several competing positions of roughly comparable philosophical plausibility. Unfortunately, there is no consensus in metaethics beyond a recognition of which metaethical positions are among the group of philosophically plausible theories, and in that way, metaethics is unlike normative ethics, where there is consensus about which specific normative theories are among the most plausible. David Bourget and David Chalmers recently conducted a survey of professional philosophers' metaethical positions, and here are some of their data:

Mean estimates for respondents who either "accept" or "lean towards" these positions. "Other" represents those respondents who selected different responses from "accept" or "lean towards" when asked to state their position on a topic.

Metaethics: moral realism 56.4%; moral anti-realism 27.7%; other 15.9%.
Metaphilosophy: naturalism 49.8%; non-naturalism 25.9%; other 24.3%.
Moral judgment: cognitivism 65.7%; non-cognitivism 17.0%; other 17.3%.
Moral motivation: internalism 34.9%; externalism 29.8%; other 35.3%
Normative ethics: deontology 25.9%; consequentialism 23.6%; virtue ethics 18.2%; other 32.3%. (Bourget and Chalmers 2014)

Subsequent analysis by Bourget and Chalmers revealed no interesting associations among the respondents' demographic backgrounds, training, geographic locations, or responses to other survey questions that are relevant to this section. It is therefore fair to take these data as a roughly accurate representation of several different fault lines of philosophical disagreement.

But what are these disagreements about? A brief explanation of the relevant metaethical positions follows. Please bear in mind that the following exegesis aims to establish a technical point: even very strong scientific naturalism is compatible with nearly all of the standard metaethical positions. With a single exception, there is no argument from ontological commitments of contemporary cognitive science either for or against any of the standard metaethical positions. It is a mistake, therefore, to conclude that, say, noncognitivism is the metaethical position that should be favored by behavioral scientists, simply in virtue of their antecedent commitment to scientific naturalism.

Realism, Naturalism, Cognitivism ...

Let us assume—purely for the sake of exegetical simplicity—that all ethical properties are either invented, intuited, or discovered. And let us also assume for the same purpose that these are mutually exclusive categories. These categories are different proposals about the nature of the relationship between the mind and ethical properties—specifically, ethical properties that are discovered are discovered because at least some ethical properties produce causal effects that, under the right conditions, can be picked up, registered, and subsequently represented by human sensory, cognitive, and social systems. Ethical properties that are intuited are done so because the mind is able to consider a range of propositions and then sort the true propositions from the false propositions. When the mind does this, the ethical properties whose existence are entailed by the truth of the relevant proposition become known, or at least represented in a person's mind. Invented ethical properties come into being because it is possible for groups of people to agree to use stipulative names for certain objects, processes, and events—in sports, names like "fair plays" and "foul balls"; in ethics, "just" and "fair"—and when a phenomenon so named is produced by a particular set of rules or procedures that are themselves subject to agreement and can be thus established by convention, there is a sense in which fair plays, foul balls, justice, and fairness come into being as by-products of an equilibrium formed out of linguistic consensus, the material properties of the things and behaviors bearing the new names, and of course various patterns of very complex collective action and feedback loops.

These three pictures of ethical properties correspond to three standard positions in metaethics: roughly the views that ethical properties are real and discovered, they are real and intuited, and they are real but only because they are invented. These positions are usually called naturalism, intuitionism, and constructivism.

Interacting with these three positions are different theories about how the mind represents ethical properties. Cognitivism says that beliefs and judgments and other mental representations can be true or false or something in between, and that beliefs and judgments are able to represent ethical properties. So cognitivism is naturally associated with naturalism and intuitionism. It is also easy to link constructivism with cognitivism. The key is to remember that constructed properties are not discovered; they are projected at first onto the world through patterns of linguistic stipulation. But nevertheless, baseballs, beaver dams, and perhaps even justice are no less real than quarks despite the fact that they are socially constructed. Noncognitivism says that beliefs and judgments and other mental representations that can be true or false do not represent ethical properties. People have beliefs and judgments that *seem* to represent ethical properties, but these beliefs and judgments are false. Ethical beliefs and judgments are caused by mental states, like some very basic emotions and feelings that lack propositional content and are therefore unable to transmit information to ethical beliefs and judgments. For that reason, noncognitivism can be linked with intuitionism. However, there are philosophers and cognitive scientists who think that emotions do sometimes convey information about the external world. If this is the case, then evidence that some moral judgments are caused by emotions is not evidence either for or against noncognitivism. Some other non-truth-functional mental process may be the sole cause of moral judgments, and maybe emotions are not a very important cause of moral judgments. This means that it is important not to confuse noncognitivism with the substantially weaker psychological hypothesis that emotions, feelings, and other affective states play a causal role in moral cognition.

Feelings are of course an important source of motivation. But that motivational force may be redundant in some cases of ethical action. Internalists hold the position that either all or some moral judgments are intrinsically motivating, so that the judgment "*x* is right" alone can be sufficient motivation to cause someone to do *x*. Externalists disagree, and hold that a moral judgment needs to link up with other motivational mental states for the judgment to cause action.

It is standard for philosophers to explain and define these positions according to how someone who holds these positions would interpret a normative statement or utterance. So, suppose that someone says in a loud and angry voice, "That's not fair!" while pointing to a particular way that the cookies have been distributed to a group of students in an undergraduate psychology seminar. Here is how a philosopher might analyze that statement:

Cognitivism: The statement expresses a judgment that is either true or false.

Noncognitivism: The statement does not express a judgment that can be true or false; the statement really expresses only that the speaker is upset.

Internalism: The judgment behind the statement itself provides motivation for the speaker to do something about how the cookies have been distributed.

Externalism: It is the upset associated with the statement that provides motivation for the speaker to do something about how the cookies have been distributed.

Naturalism: The speaker knew that the distribution was unfair because she applied knowledge that was previously acquired (not necessarily by her) partly in virtue of the causal effects of either unfairness or some other ethical properties on human cognition.

Constructivism: The speaker is right to call that distribution unfair because disinterested rational agents would see the distribution the same way.

Intuitionism: The proposition "distributing goods in such-and-such a way" is true and known a priori by way of the speaker's intuition; her statement reflects a straightforward inference from this proposition and the observation that her intuition has been violated.

So a cognitivist-externalist-constructivist might interpret the statement as true because the property the statement is about is invented, and the motivation to redistribute the cookies is rational because the emotions behind it are in harmony with this judgment. It is left as an exercise for readers to work out the interpretations that flow from the other combinations of metaethical positions.

Let us make one more simplifying assumption: none of these positions can be synthesized, and so constructivism-intuitionism is not a possible metaethical position. Since noncognitivism does not go together with naturalism, constructivism, or intuitionism, this means that there are eight different metaethical positions that can be adopted. These eight positions are a good approximation of the positions that are treated as the most philosophically plausible in contemporary metaethics—although the truth is that the ratio of published metaethical positions to published metaethicists is roughly 1:1.

Suppose now that we design a scientific experiment that tests a hypothesis about moral cognition. Data are collected and analyzed; the main effect is huge, so the result is then written up and a paper sent off to a journal like *Psychological Review*. Will relying upon any of the eight metaethical positions send up red flags for potential reviewers of the paper? That is, from the perspective of an expert in cognitive science or psychology, are any of these metaethical positions scientifically implausible and thus not fit for inclusion in a scientific paper? This question provides us with an easy way of seeing how judgments of philosophical plausibility and scientific projectability can come apart. For the only metaethical position that fails this test—and so would be inappropriate to use as the basis for a control condition in an experiment, because it may impair the reliability of the experiment in the eyes of reviewers—is intuitionism. It is not part of the standard metaphysics or epistemology of cognitive science that the mind is able to know a priori truths through a simple act of intuition-based inference.

We need to be careful, however. The claim here is not that intuitions are not part of the ontology of cognitive science. They are. It is, instead, the philosophical proposition that the mind can intuit its way to knowledge that is problematic from the perspective of cognitive science. The grounds of cognitive science's rejection of the philosophical claim are metaphysical. It is very nearly an axiom of cognitive science that there is no way that the mind can access the objects and properties about which it thinks except by causal interactions of various different forms and durations with the external world. Of course, this is not to assert that the only causal interactions that are relevant occur within the human lifespan, and to thereby deny the existence of innate ideas or core knowledge. The claim is that it is only through causal interactions between external phenomena and either innate or acquired ideas that the mind develops knowledge. So cognitive science's disagreement with Ross's view that "both in mathematics and in ethics we have certain crystal-clear intuitions from which we build up all that we know about the nature of numbers and the nature of duty" is not that there are no intuitions but, instead, with the specific thesis that some intuitions can produce knowledge in a way that is completely independent of the causal influence of the world on the mind (see Bengson 2015 for a modern development of this idea). What is denied by cognitive science is the possibility of a kind of noncausal relation between intuitive mental experiences in the mind and the objects of those experiences, not the reality of the intuitions themselves.

So intuitionism does not seem to fit with current judgments of scientific plausibility; only intuitionism is incompatible with even a very strong form of scientific naturalism. But this shows us that, more importantly, all of the remaining metaethical positions remain compatible with the metaphysics of contemporary cognitive science. We find, then, that judgments about the scientific plausibility of the standard metaethical positions are only slightly more restrictive than judgments of philosophical plausibility about the same. For philosophical intuitionism remains credible among contemporary ethical theorists (see Huemer 2005; Audi 2004; Stratton-Lake 2002), and the scientific intuitionism of moral psychologists like Jonathan Haidt is really a form of noncognitivism (Clarke 2008).

A Third Methodological Principle

We now reach the third and final of this chapter's methodological lessons.

3. *Research in moral psychology that has implications for the truth of any particular metaethical position is biased if its protocols and methodology reflect a presumption either in favor of the relevant metaethical position or against any of the other philosophically plausible metaethical theories. Therefore, unless there are controls for at least some—and ideally all—relevant and independently plausible metaethical positions, research in moral psychology should remain neutral with respect to substantial metaethical questions.* This principle is actually the application of a general rule of experimental science to the intersection of metaethics and moral psychology. A scientific experiment that is meant to test hypothesis H and yet uses, for example, dependent variables that are biased against a set of scientifically plausible alternative hypotheses H_1, H_2, and H_3 will be unable to confirm H or to refute H_1, H_2, or H_3. The reason is that any data produced by the experiment are not logically independent of the truth of H, or the falsity of the alternative scientifically plausible hypotheses. It thus could be the design of the experiment, and not the data, that is the basis of any conclusion favoring H. Likewise, a scientific experiment that is designed by researchers who are unaware of the scientifically plausible alternatives to their favored hypotheses H will only by sheer metaphysical luck be a fair test of H. Yet as we will see in chapter 5, it is easy to find experiments in moral psychology that do not accurately control for the fact that there are multiple projectable metaethical positions.

But before moving on to the examples that demonstrate that point, it is helpful to take explicit note of the fact that intuitionism, naturalism, and constructivism are not psychological hypotheses, and neither are they hypotheses that can be confirmed or refuted by the methods of experimental psychology. They are, instead, metaphysical hypotheses about the relationship between ethical properties and human cognition. So, empirical evidence that intuition plays a role in moral cognition does not establish the truth of intuitionism because it does not establish that either naturalism or constructivism is false. Likewise, evidence that there are constructivist elements in moral thought does not establish the truth of constructivism because it does not establish that either naturalism or intuitionism is false. But—and this is the key point—determining that naturalism is false requires historical evidence that all meaningful attempts to acquire knowledge through causal interactions with ethical properties have led to failure, and even the very best psychological experiment would be unable to produce such data.

Perhaps the easiest way to appreciate that conclusion is to use a direct analogy with naturalism about scientific theories in biology, chemistry, physics, or, indeed, any other area of science that is organized around a core of mature, canonical theory. It is a fact about the psychology of scientific reasoning, scientific concept and theory formation, and scientific concept and theory revision that there are intuitive and constructivist elements of all of these processes (Medawar 2013; Nersessian 1984, 2008; Koslowski 1996). Yet it does not follow from this that either constructivism or intuitionism about, say, chemistry is true. Evidence that the properties posited by chemical theory exert no causal influence on either the human minds or the social practices of science would be historical, since the best evidence of the relevant influence (or lack thereof) of, for example, protons and kinetic isotope effects on cognition is chemistry's increasing ability over generations of research to achieve some of its original epistemic ends (or its failure to do so). The truth of naturalism, constructivism, or intuitionism in chemistry cannot be settled by studying only the minds of chemists; the truth of ethical naturalism, constructivism, or intuitionism cannot be settled by moral psychology alone.

References

Audi, R. 2004. *The Good in the Right: A Theory of Intuition and Intrinsic Value*. Princeton, NJ: Princeton University Press.

Bengson, John. 2015. "The Intellectual Given." *Mind, a Quarterly Review of Psychology and Philosophy* 124 (495): 707–760.

Bourget, David, and David J. Chalmers. 2014. "What Do Philosophers Believe?" *Philosophical Studies* 170 (3): 465–500.

Clarke, Steve. 2008. "SIM and the City: Rationalism in Psychology and Philosophy and Haidt's Account of Moral Judgment." *Philosophical Psychology* 21 (6): 799–820.

Huemer, Michael. 2005. *Ethical Intuitionism*. New York: Palgrave Macmillan.

Koslowski, B. 1996. *Theory and Evidence: The Development of Scientific Reasoning*. Cambridge, MA: MIT Press.

Medawar, P. B. 2013. *Induction and Intuition in Scientific Thought. Routledge Library Editions: History & Philosophy of Science*. New York: Taylor & Francis.

Mill, J. S. 1925. *Utilitarianism*. Raleigh NC: Hayes Barton Press.

Nersessian, Nancy J. 1984. *Faraday to Einstein: Constructing Meaning in Scientific Theories. Martinus Nijhoff Philosophy Library*. London: Springer.

Nersessian, Nancy J. 2008. *Creating Scientific Concepts*. Cambridge, MA: MIT Press.

Shanley, Mary Lyndon, and Carole Pateman. 1991. *Feminist Interpretations and Political Theory*. New York: Wiley.

Stratton-Lake, Phillip. 2002. *Ethical Intuitionism: Re-evaluations*. Oxford: Oxford University Press.

5 Inconsilience between ADJDM Moral Psychology and Philosophical Ethics

In contemporary moral psychology, some of the most popular experimental protocols, and some of the equally popular theories that these protocols have been used to develop, violate the three methodological principles derived in chapter 4; the implication being that moral psychology and philosophical ethics are thereby at risk of inconsilience. Providing evidence of this inconsilience is one of the goals of this chapter. But it is also important for us to examine the cause the emerging inconsilience—because with an understanding of its underlying source we can develop a realistic solution to it. For that reason, the second goal of this chapter is provide an explanation of the emerging gap between moral psychology and philosophical ethics, and the explanation that I will offer is one in which there are neither moral villains nor failures of scientific practice. If anything, the inconsilience is the unfortunate by-product of an interaction between a technical property that is unique to theories in philosophical ethics and two basic and otherwise entirely rational methodological principles of scientific research.

The first of these two basic principles concerns external validity, which, informally, is the idea that no psychological theory should be regarded as confirmed if, over the medium term, it is impossible to find evidence that the theory describes, to at least a good approximation, some psychological mechanism or process that can occur outside of a lab. The second basic principle holds that *all* science relies as much as possible upon concepts and methods developed by previous generations of research in the relevant field—and this principle is reliable because it is only the approximate truth of this past research that can explain the observation that, historically, science has made (incremental and nonuniform) progress (Boyd 1990, 1991). According to the second principle, what it is for a theory to be projectable in the sense given to that concept by Goodman, and therefore apt for induction and explanation, is merely for the theory to be plausible in light of previous scientific research.

Now, these two principles—we can call them the principles of *external validity* and *projectability*—are embedded in the practice of most contemporary work in moral psychology, just as they can be found in most other fields of scientific research. But unfortunately, the metaphysics of the relationship between philosophical ethics and moral psychology are such that a moral psychologist who is operating according to these methodological principles will eventually be induced to accept theoretical conclusions that are deeply implausible from the perspective of philosophical ethics. The evidence for this is most easily presented as the following inferential chain:

Theories in psychological science must be externally valid → The core normative concepts of philosophical ethics do not refer to any actual patterns of thought or patterns of social practice (that is the property unique to theories in philosophical ethics, namely, that they are *unrealized*) → So, trivially, it will not be possible to produce both externally valid theories in moral psychology that also employ *only* the core concepts of philosophical ethics → But projectability requires that moral psychologists integrate as much as possible the core concepts of philosophical ethics into scientific practice → The core concepts of philosophical ethics must therefore be redefined, and done so in ways that allows for meaningful tests of the external validity of theories in moral psychology.

And so, the thesis of this chapter is that this chain of inferences represents something deep about the practice of a large number of contemporary moral psychologists—and that, crucially, the resulting inconsilience between moral psychology and philosophical ethics is emerging *because* the redefinitions (and associated methodological practices) have become too severe.

But again, I stress that my aim here is not polemical, for another way of putting this chapter's thesis is that the interface between philosophical ethics and moral psychology is a place where ordinarily reliable methodological principles can break down unless we are extremely clear about the metaphysics of the relationship between ethics and moral psychology. (The reliability of methodological principles in the behavioral sciences therefore is context dependent.) And the solution that I will offer—developed over the course of chapters 7 through 10—rejects the idea that projectability between philosophical ethics and moral psychology requires that moral psychologists directly import the core concepts of philosophical ethics into scientific practice. To understand this alternative, it is helpful to resurface the idea of conceptual framework pluralism—which, please recall, is that idea that the best way to understand a complex system (like normative behavior in humans) is to develop a plurality of different conceptual vocabularies useful for generating useful models, frameworks, and theories, and to develop in parallel a corresponding "theoretical perspective," which is a metaphysical theory that represents how each of the different conceptual vocabularies is related to one another. The proposed solution for the problem of inconsilience, then, involves: (1) adopting a methodological framework that holds that philosophical ethics and moral psychology should use substantially different conceptual vocabularies; (2) developing a theory that tells us which different and mostly nonoverlapping metaphysical dimensions of human normative behavior moral psychology and philosophical ethics are, respectively, about; and (3) showing that the conceptual and methodological independence of philosophical ethics and moral psychology can be a source of novel research questions for moral psychology. Crucially, this alternative theoretical path frees moral psychology from the need to redefine normative concepts from philosophical ethics so that they refer to measurable psychological phenomena.

Let us now turn our attention to the defining characteristics of the research in moral psychology that often violates the methodological principles derived in chapter 4, and while doing this, I will clarify what it means for theories in moral psychology to have external validity. We will then turn to some of the evidence which indicates that different research programs in moral psychology violate the three methodological principles introduced in chapter 4.

Moral Psychology as ADJDM

The most commonly used method in contemporary moral psychology consists of asking people—usually undergraduates who have volunteered to participate in a small study—for judgments about hypothetical moral dilemmas. Conclusions about how the mind implements moral cognition are then extrapolated from any patterns in the sampled judgments that can be associated with patterns in the details of the dilemmas themselves. This research is typically experimental, if only in the sense that participants in the relevant studies are randomly assigned to different conditions, and that data is generated by comparing the responses of participants across conditions. The hypothetical scenarios usually function as the independent variable for each condition, and judgments about these scenarios thus are the dependent variables. The judgments are also standardly called "moral intuitions" because subjects are typically not asked to explicitly reason about the dilemmas, and neither are they given the social (friends, experts) and material (books, Internet connections) resources that support sustained social deliberation. After this point, it will be helpful to have a name for the specific kind of research that I am talking about, and the name "trolleyology" has been proposed in conversation by a number of people. But that is not the name I would like to use, because it suggests that the only stimuli used in the relevant studies are variations of the trolley problem (an explanation follows), which is false. A broader name is more appropriate, and "adult deontic judgment and decision making" (ADJDM) both fulfils this requirement and helpfully calls attention the similarities between moral psychology and the methodology in the larger field of judgment and decision making (for examples, see Gilovich and Griffin 2010; Connolly and Ordóñez 2003).

That said, the paradigmatic scenario in ADJDM-style moral psychology is some version of Philippa Foot's famous trolley problem (Foot 2002; Thomson 1976). An experimental participant reads a short story in which a runaway train races down some tracks toward a small group of people trapped and unable to escape. A bystander sees the runaway train, and also sees that she has the opportunity to divert the train onto a side track, where only a single person is trapped. The bystander can decide not to intervene and five people will die, or decide to intervene by pulling a switch to divert the train, and then only one person will die. Participants in these studies are often asked what the bystander should decide to do, and how right or wrong the bystander's decision is. More often than not, participants are asked to make their judgments by taking the perspective of the bystander. These experiments often use a standard two-alternative, forced-choice probe to record participants' responses; in another version the scenario simply describes an either/or choice that the experimental participant is asked to evaluate according a number of different scales and measures. And while there are many, many variations (see, e.g., Lanteri, Chelini, and Rizzello 2008; Rai and Holyoak 2010; Bleske-Rechek et al. 2010; Di Nucci 2013; Navarrete et al. 2012; Swann et al. 2010; Mikhail 2011; Duke and Bègue 2015; and Royzman, Landy, and Leeman 2015), what nearly all have in common is that they require subjects to decide between two mutually exclusive actions, and this decision is treated as a paradigmatic moral judgment. Stijn Bruers and Johan Braeckman recently reviewed the psychological and philosophical literatures on judgments about trolley problems (Bruers and Braeckman 2014). They provide a helpful breakdown of the different variations of both the trolley problem itself, and more importantly, they find no evidence of a consensus interpretation of the judgments elicited by the use of these dilemmas over the last thirty-five years. Instead, they report at least six different "algorithmic" and seven different psychological explanations of patterns of judgments about trolley problems that appear in the literature.

But again, ADJDM-style moral psychology is not precisely coextensive with trolley-problem experiments. ADJDM-style research extends to the use of other hypothetical dilemmas to probe for patterns in participants' moral judgments, and, generally speaking, this approach to studying moral cognition has some fairly obvious scientific virtues. Data collection is very cheap, as participants can either be interviewed fairly quickly or their judgments can be collected using online surveys. Marc Hauser and his coauthors write: "the use of artificial moral dilemmas to explore our moral psychology is like the use of theoretical or statistical models with different parameters; parameters can be added or subtracted in order to determine which parameters contribute most significantly to the output" (Hauser et al. 2007, 4). What's more, no particularly advanced forms of training in either statistics or the methodology of behavioral research are required to run ADJDM-style experiments. All that is necessary is the creativity to formulate new hypothetical dilemmas, access to willing participants, an ability to survey and record their judgments, and enough statistical know-how to use null-hypothesis tests to find patterns of the responses of judgments. ADJDM-style moral psychology's clear value proposition surely explains its popularity among psychologists and philosophers.

External Validity

How can we tell, then, whether or not ADJDM-style experiments have external validity? To answer this question, we need to distinguish between universal external validity and local external validity. If some psychological mechanism PM is posited upon the basis of a set of experimental findings EF, then the hypothesis that PM exists has *universal validity* only if approximately the same experimental protocols as those that were used to produce

EF would produce approximately the same data as EF when the protocols are applied to participants who are sampled from *any* human population whatsoever. By contrast, if some psychological mechanism PM is posited on the basis of a set of experimental findings EF, then the hypothesis that PM exists has *local validity* only if approximately the same experimental protocols as those that were used to produce EF would produce approximately the same data as EF if and when the protocols are applied to participants who are sampled from *some* (and not necessarily any) human population, and where this second population is known to be a different population than that of the original experiment. So, an experiment that confirms the existence of some universal properties of the human mind will have universal validity; experimental findings about, for example, whether liberals or conservatives in America prefer nouns (Cichocka et al. 2016) will have local validity only.

Now, one of the things that we will see over the course of this section and the next is that both critics and practitioners of ADJDM-style moral psychology have taken positions that reflect a shared assumption that universal validity is the appropriate standard for moral psychology to aim for. And almost immediately, there are a number of common-sense objections to such a methodological view. For example, high-stakes decisions like the trolley dilemma are relatively uncommon in most people's lives, and nearly every moral problem that people confront has a more complicated decision structure than that of a strict either/or choice that either can be or must be decided without any planning or forethought, let alone further research and deliberation. Most real moral problems can only be solved by making a sequence of choices, and finding the resolution involves knowledge of local norms and customs as well as the use of social behaviors like interpersonal deliberation and reference to the knowledge of experts (Rai and Fiske 2011). So, we already know that the protocols used in ADJDM are not perfectly representative of the real diversity of moral judgments and moral problems.

But let us pass by that issue to consider some deeper objections to the universal validity of ADJDM-style moral psychology. ADJDM dilemmas typically ask participants to decide if one course of action is permissible; as Richard Shweder observes, this generates the following conceptual problem: "the English word 'permissible' might mean any of the following things: 'Permitted by law,' 'Not punishable by either god, nature, conscience or other human beings,' 'Not damaging to your reputation,' 'Not morally incorrect,' 'Likely to be viewed with approval by god, nature, conscience or other human beings'" (Shweder 2014, 275). And then, Shweder goes on to note that the trolley problem itself resembles rather strikingly a famous Oriyan myth. In this myth, there is a father who is "trapped" in a position analogous to that of the bystander in the original trolley problem: he must make a choice that determines whether a proportionally larger or smaller number of people die. The details matter, so permit me a little bit of exegesis. A king has ordered that a temple be constructed in exactly twelve years, and that the work be done by no more and no less than 1,200 workers, and so done under the direction of a master architect, who is also the father I have just mentioned. If these conditions are not met, then the king will kill all the workers. And so, as the temple nears completion and time is running out, there arises the difficult problem of how to place the capstone. The master architect cannot solve the problem but, completely unbeknown to the architect, his twelve-year-old son, who is an architecture prodigy, comprehends perfectly the problem of placing the capstone and designs a solution. Without the father's knowledge of who designed the capstone, the stone is shaped and placed, and the temple finished—although, because of the boy's contribution, the temple has been built using 1,201 workers. The workers plead with the master architect to kill the additional laborer and appease the king, thereby sparing their lives. They protest to the master architect, "It is a simple calculation: kill one to save more than a thousand!" The master architect relents, and seeks out the mysterious architect—who he then sees for the first time is his own son. He cannot carry out the murder. Soon after, the boy learns of his father's choice.

Shweder remarks: "Over the years, I have discovered that most of my American students are unable to predict the conclusion of the Oriya legend on the basis of the plot summary" (Shweder 2014). How does the story end? The boy, recognizing his father's moral paralysis, climbs to the top of the temple and throws himself off, sacrificing himself. But he doesn't do this to save the lives of the workers. Instead, the boy kills himself to save his father from a situation in which his only two options are both deeply sinful. The point is that, in real life, there is always a third choice; and lest Shweder's example seem too apocryphal, among the Iglulik Inuit it was customary for elders to request that they be killed—and usually violently, on the belief that such a death aided the soul in its passage to a better world—if they had become burdensome to young members of their community (Kirmayer, Fletcher, and Boothroyd 1998).

So, if ADJDM-style moral psychology aims for universal validity, it will almost surely fail. Local external validity is a more reasonable goal. But then, this requires something that has not been developed within either moral psychology or philosophical ethics: a social theory that is both philosophically plausible and useful for determining which specific normatively integrated populations are the appropriate ones to sample from in order to test the local validity of any given theory in moral psychology. For, without a theory of which social behaviors are *moral* or *ethical* behaviors, it will be impossible to tell if the initial tests sampled from a population that routinely manifests ethical or moral behavior. Note, however, that a socially realistic theory of ethical behavior is not among the needs of researchers who are striving for universal validity, since any population whatsoever will do as the independent population to sample from.

Now, it appears that there are two different strategies in contemporary ADJDM-style moral psychology for avoiding failures of universal external validity. The first strategy is to pursue a very strong form of reductive methodological individualism about morality. This involves attempting to define moral cognition in terms that refer only to the operations of very low-level features of the mind-brain system—features that could be developmental universals of human psychology. Proponents of this approach typically associate it with evolutionary theory, most often through the use of predictive inferences from ultimate scenarios about the evolution of morality. However, please recall one of the lessons from chapter 2, that these inferences are fallacious (because they violate Mayr's lemma), and so it would be a mistake to think the reductive approach to moral cognition is compatible with evolutionary biology. The second strategy is to adopt a theory of ethical behavior that is so utterly trivial that it cannot possibly fail to be realized by almost any human population. We will encounter examples of these two strategies presently, and unfortunately, both of these strategies lead to inconsilience with philosophical ethics.

ADJDM-style Moral Psychology Is Philosophically Implausible

Earlier we arrived at three methodological lessons for moral psychology, which, reworded for concision, are:

1. It is a category mistake to try to investigate the psychological representations or physical presuppositions of families of ethical theory. In order to avoid this mistake, researchers should have reasonably accurate knowledge of the structural content of the regulative framework whose psychological representation or presuppositions they are investigating.
2. Moral psychology should only investigate the psychological foundations of the most plausible theories in philosophical ethics.
3. Research in moral psychology that has implications for the truth of any particular metaethical position is biased if its protocols and methodology reflect a presumption either in favor of the relevant metaethical position or against any of the other philosophically plausible metaethical theories. Therefore, research in moral psychology should remain neutral with respect to substantial metaethical questions.

The remainder of this chapter provides a catalog of some of the characteristic ways that ADJDM-style moral psychology violates these methodological principles.

Deontology and Consequentialism Are Not Natural Kinds in Psychology

The most common way in which ADJDM-style moral psychology falls out of compatibility with philosophical ethics is for researchers in moral psychology to define consequentialism and deontology as psychological categories or kinds, and thus depart rather radically from the historically accurate view that consequentialism and deontology are two families of normative theory in philosophical ethics. Recall our earlier mention of Joshua Greene's view that "that the terms 'deontology' and 'consequentialism' refer to psychological natural kinds" (J. D. Greene 2010). Of course, the motivation for this recategorization is often to license the practice of interpreting data that represents judgments about hypothetical moral dilemmas as indicating whether or not "people" (undergraduates mostly; see Henrich, Heine, and Norenzayan 2010 for a discussion of related issues) are consequentialists or deontologists. For an example that we will return to in more detail in the next chapter, Marc Hauser and his colleagues coded as inadequate the responses of subjects who, to justify their judgment about one trolley dilemma, used "utilitarian reasoning (maximizing the greater good) and [who justified] their judgment of the other [trolley dilemma] using deontological reasoning (acts can be objectively identified as good or bad) without resolving their conflicting responses" (Hauser et al. 2007,

14). The point is that utilitarianism and deontology are treated as mutually exclusive and mutually incompatible styles of reasoning.

Please keep in mind that I am not arguing that the constructs *deontological judgments* and *consequentialist judgments* are false or misleading; my complaint is subtler: there is no reason to think that these two specific constructs have anything to do with the most plausible consequentialist and deontological theories in philosophical ethics. To the extent that these constructs pick out real patterns of normative judgment and justification, they could just as meaningfully be called *pattern-a judgments* and *pattern-b judgments*. But as we have seen, defining utilitarianism, consequentialism, and deontology—or, indeed, any family of philosophically plausible ethical theories of which posits a distinctive regulative framework—as a natural kind of psychology is a category mistake. Deontology and consequentialism are families of different normative ethical theories, and theories in both families are individuated by the structural content of their regulative frameworks. As we noted earlier, it is not possible to ask a group of experimental participants to concurrently realize Mill's utilitarianism, Sidgwick's utilitarianism, and Singer's utilitarianism—which means that there will not be a psychological signature that is common to the realization of each member of a families of plausible ethical theory. As we noted earlier, asking a group of people to concurrently implement Mill's, Sidgwick's, and Singer's utilitarianism would be like asking a group of people to concurrently play Olympic rules hockey, NHL rules hockey, and curling. Nor is it possible to study the psychological representations associated with these three utilitarian theories by studying hypothetical judgments that involve computing some kind of abstract moral trade-off. That would be analogous to proposing that it would be possible to study psychological representations of winter sports by conducting a study that required participants to make judgments about the relative velocity of a dark obloid object moving along a flat plane, and where the color of the plane contrasted with the color of the moving object. The problem in both cases is that constructs like winter sports (qua family of discrete sports) and utilitarianism (qua family of discrete possible regulative frameworks) simply do not have distinctive psychological representations.

Here are some examples of this category mistake. Peter DeScioli and Robert Kurzban have set themselves the task of "solving the mysteries of morality" (DeScioli and Kurzban 2009), which for them is the task of producing a psychological explanation of patterns of judgments about trolley problems that is also plausible in light of extrapolative inferences from a number of evolutionary scenarios. They write, "'Morality' is an umbrella term for a collection of different psychological systems." Moral dilemmas emerge when "consequentialist" decision making conflicts with "Kantian" or "deontological" decision making. These patterns of decision making are given technical definitions by DeScioli and Kurzban:

> An outcome consists of payoff to the organism and other relevant organisms, i.e., a vector **y** of payoffs to the self and others ... the decision maker has a standard utility function for ranking outcomes depending on the resulting payoffs $u(\mathbf{y})$.
>
> A consequentialist decision procedure would choose an action, a^*, to maximize utility:
>
> max $u(\mathbf{y})$, (1)
>
> $a \epsilon A$
>
> [...] In contrast, the Kantian decision procedure would choose an action not only based on the payoffs **y**, but also based on whether the action is labeled morally wrong ...
>
> max $u(\mathbf{y})$, subject to the constraint, $a \notin W$ (2)
>
> $a \epsilon A$
>
> (Kurzban, DeScioli, and Fein 2012, 2)

They continue, "Moral dilemmas occur when maximization based on payoffs (1) conflict with moral constraints (2) ... [and that] ... these conflicting decision procedures are used to some extent, which is why humans perceive [trolley problems] as 'dilemmas' rather than having clear-cut solution." Thus, DeScioli and Kurzban use "consequentialist" and "Kantian" to refer to two different reasoning strategies, not to families of normative theory. As noted earlier, these two strategies could just as informatively be called *pattern-a* and *pattern-b*, and using those names would actually be an improvement, since the new names would no longer create the false impression that the two strategies defined by DeScioli and Kurzban have some meaningful connection with ethics.

Twice now we have mentioned Joshua Greene's view that "utilitarianism" and "deontology" refer to psychological kinds, so let us now take a closer look at his work. Greene has collaborated with a number of researchers over the last decade, and these collaborations have produced some of the most influential research in ADJDM. Greene and his coauthors believe that philosophical ethics is not a rational practice that has made some progress toward, for example, discovering which regulative frameworks are most likely to be ethical regulative frameworks. Instead, their view seems to be that philosophical ethics is really just the historical output of a collection of mostly well-off American and European men whose defining cultural practice seems to be to use articles, lectures, and books to document their history of working through a tension between two species-typical, developmentally endogenous modes of using deontic concepts:

> We propose that the tension between the utilitarian and deontological perspectives in moral philosophy reflects a more fundamental tension arising from the structure of the human brain. The social-emotional responses that we've inherited from our primate ancestors ... shaped and refined by culture bound experience, undergird the absolute prohibitions that are central to deontology. In contrast, the "moral calculus" that defines utilitarianism is made possible by more recently evolved structures in the frontal lobes that support abstract thinking and high-level-cognitive thinking. ... Should this account prove correct [despite its speculative nature] it will have the ironic implication that the Kantian, "rationalist" approach to moral philosophy is, psychologically speaking, grounded not in principles of pure practical reason, but in a set of emotional responses that are subsequently rationalized. (Greene et al. 2004, 398)

What kind of psychology finds expression in the utilitarian and deontological perspectives? Greene and his coauthors define "utilitarianism" and "deontology" ("non-utilitarianism") as mutually exclusive decision-making processes: "To test the hypothesis that utilitarian moral judgments engage brain areas associated with 'cognitive' processes, we compared the neural activity associated with utilitarian judgments (accepting a personal moral violation in favor of a greater good) to non-utilitarian judgments (prohibiting a personal moral violation despite its utilitarian value)" (Greene et al. 2004, 392). Greene and colleagues reported a difference in neural activation, and interpreted this as strong support for their hypotheses. However, Julia Christensen and Antoni Gomila (Christensen and Gomila 2012) have argued that this difference can also be explained by a number of variables that go beyond the purported neuroethical processes causing the judgment—factors such as the number of words in the stimulus scenario and incidental features of both the stimuli scenario itself and the experimental constructs. More work is necessary to decisively confirm or refute Greene and his collaborators; but the outcome of this further research cannot change the fact that these researchers are making the same category mistake as Kurzban and DeScioli. Again, no scientific information is lost, and an appreciable amount confusion is avoided, if the constructs utilitarian and nonutilitarian judgment are renamed to avoid deliberately suggesting that they have anything to do with progress in moral philosophy.

The research of Eric Uhlmann, David Pizarro, David Tannenbaum, and Pater Ditto provides us with our final and most sophisticated example of the redefinition of utilitarianism and deontology so that these concepts refer to psychological kinds. Once again the two families of ethical theory are reconceptualized as two different cognitive strategies for making decisions (Uhlmann et al. 2009; Ditto, Pizarro, and Tannenbaum 2009). Pizzaro and his coauthors constructed a scale that provides an operational measure of whether someone is a utilitarian or a deontologist in terms of an experimental participant's preferences for solving abstract dilemmas. Here is their scale:

(1) "Is sacrificing Chip/Tyrone to save the 100 members of the Harlem Jazz Orchestra/New York Philharmonic justified or unjustified?"
(1 = completely unjustified, 7 = completely justified)
(2) "Is sacrificing Chip/Tyrone to save the 100 members of the Harlem Jazz Orchestra/New York Philharmonic moral or immoral?"
(1 = extremely immoral, 7 = extremely moral)
(3) "It is sometimes necessary to allow the death of an innocent person in order to save a larger number of innocent people."
(1 = completely disagree, 7 = completely agree)
(4) "We should never violate certain core principles, such as the principle of not killing innocent others, even if in the end the net result is better."
(reverse coded; 1 = completely disagree, 7 = completely agree)
(5) "It is sometimes necessary to allow the death of a small number of innocents in order to promote a greater good."
(1 = completely disagree, 7 = completely agree)
(Uhlmann et al. 2009, 482)

The idea here is clear. A participant is *defined* as a consequentialist if her preferences tend closer to 7, and *defined* as deontologist if her preferences tend closer to 1. And while the data in their study does show, among other things, that subjects who self-reported their political orientation as liberal tended away from 1 when the prompt used the name "Tyron" and tended closer to 1 when the prompt used the name "Chip," these data do not support such deep conclusions as: "We found that liberals were less willing to endorse the killing of an innocent person on consequentialist grounds when the name of the individual suggested he was Black than when it suggested he was White." The technical reason why such data do not support this conclusion is that "consequentialist grounds" properly refers to actual consequentialist ethical theory about killing innocents, as discussed in, for example, Scheffler 1988; Williams and Caldwell 2006; and Adams 1999. The phrase "consequentialist" could be replaced with a phrase like "participants who tended to express judgments closer to 1 on our scale" without any loss of meaning or scientific content.

More interesting, though, is the profoundly deep conclusion about moral cognition that Pizarro and Tannenbaum think follows from the data described earlier:

> Rather than being moral rationalists who reason from general principle to specific judgment, it appears as if people have a "moral toolbox" available to them where they selectively draw upon arguments that help them build support for their moral intuitions. While the present studies do not imply that general principles never play a direct, *a priori* role in moral judgment, they do suggest that moral judgments *can be* influenced by social desires or motivations, and that moral principles *can be* rationalizations for other causes of the judgment. Sometimes principles are used in an impartial manner. The disturbing implication of recent research, however, is that we have no easy way of verifying whether the principles we passionately invoke for our moral judgments are truly guiding our judgment in the unbiased fashion we think they are. (Pizarro and Tannenbaum 2009, 498; emphasis in original)

However, Pizarro and Tannenbaum cannot reach their first conclusion because, in the context of philosophical ethics, it implausible to conclude that preferences that tend toward 1 are literally definitive of consequentialist principles and preferences that tend toward 7 are definitive of deontological principles. That is simply because there are, for example, philosophically plausible consequentialist ethical theories that recommend attitudes that align with one side of the scale, while others will recommend attitudes that align with the other side (see Albee 2014). Because consequentialism and deontology are families of normative theory, there is no unique set of beliefs or decision-making strategies that are definitive of each family. It is a mistake to construct an operational measure for the psychological representations of consequentialism and deontology as such.

We can also take note of the fact that Pizarro and Tannenbaum need the first conclusion about "consequentialist grounds" to reach their second conclusion—again, that we can never tell when moral principles are post hoc rationalizations of, rather than causes of, moral judgments. (But does it really matter? Are moral judgments *only* reliable if they are immediately caused by conscious representations of moral principles, as opposed to simply being consistent with or entailed by well-justified moral principles? Note that the latter relation is not sensitive to either the temporal or the causal order in which representations of the principles and judgments occur in a person's mind.) Their inference to the skeptical conclusion only works if, as a conceptual truth, moral principles cannot be sensitive to racial or proprietary-to-America political properties. And while that is true of many ethical principles discussed in philosophical ethics, since racism continues to be a very real problem in the populations studied by Pizarro and his fellow researchers, it is clearly false that people's actual *realized* ethical views are not sensitive to racial or political properties. Pizarro and Tannenbaum are therefore conflating studying idealized but unrealized norms in philosophical ethics with the psychological project of describing how ethical knowledge is represented in the minds of people. Or to put the same conclusion in more straightforward language: a more cautious reading of the data is that it indicates only that their participants' actual ethical attitudes are not realizations of any of the regulative frameworks posited by philosophical ethics.

The three examples of ADJDM research in moral psychology are not the only examples of research that redefines consequentialism and deontology into concepts that refer to very simple patterns of reasoning and decision making. For some further illustrative examples, see Conway and Gawronski 2013; Körner and Volk 2014; Bartels and Pizarro 2011; Royzman, Landy, and Leeman 2015; Lee and Gino 2015.

Ad Hoc Normative Theories in ADJDM

While we have established that it is a mistake from the standpoint of philosophical ethics to define consequentialism and deontology as psychological kinds, we have not yet considered the possibility that the research programs described are investigating truly novel ethical theories—theories that are, despite their names, intended to be more psychologically realistic alternatives to many of the mainstream theory in philosophical ethics. According to this line of reasoning, no category mistakes are really being committed because Kurzban and DeScioli, for example, are most reasonably interpreted as putting forward "max $u(\mathbf{y})$, subject to the constraint, $a \notin W$"—where a is an act someone can choose, W a set of actions that cannot be chosen, and y an ordering of outcomes according to utility function, u—as a genuine ethical theory, as a genuine alternative to, say, Mill's utilitarianism or Kant's ethics. In the terminology of this book, "max u(y), subject to the constraint, a$\notin W$" would be a formula for generating the structural content of the ethical regulative framework. Note furthermore that, since these putative ethical theories are either simple decision strategies or small sets of ordered preferences about straightforward principles, they will almost automatically have an appreciable degree of universal external validity. Given the thinness and ease of realizing the relevant theories, it will not be difficult to find evidence of that realization in almost any social world.

But there is an immediate problem with this line of reasoning. If the decision strategies and processes we reviewed in the preceding section are interpreted as ethical theories, then it is not hard to show that they are deeply implausible from the perspective of philosophical ethics. The most direct way to show this is to note that, when interpreted as candidate consequentialist or deontological ethical theories, each theory fails a test of evidential parity with the best available examples in normative ethics. For, the justification proffered for these ethical theories *qua* ethical theories is somewhere between scant and nonexistent; it consists almost entirely of evidence that people's judgments about hypothetical cases correspond more or less to the structural content of the relevant ethical theory. But the fact that some experimental participants have judgments about hypothetical moral scenarios that are consistent with a principle like "max u(y), subject to the constraint, a$\notin W$" is not evidence for the conclusion that the principle itself is an *ethical* principle. That this inference is unsound is easiest to see by using analogy: that some people have played games of hockey using field-hockey sticks is not evidence that field-hockey sticks should be used in games of hockey. More generally: purely descriptive evidence that a certain pattern of thought or behavior can or does occur implies no interesting normative conclusions; that is true even when the relevant patterns of thought and behavior cannot be described except by using normative concepts. Normative premises are required to reach normative conclusions; generating the kind of evidence appropriate for justifying relatively deep normative premises and conclusions is of course one of the characteristic activities, and epistemic ends, of philosophical ethics. But none of the evidence for the decision strategies or preference sets discussed here contains any independently plausible normative propositions that can justify the hypothesis that, for instance, "max u(y), subject to the constraint, a$\notin W$" is an ethical principle. If the relevant principles or are real ethical theories, then they are profoundly underdeveloped. It counts, of course, that the theories have a modicum of psychological tractability, but it does not count for very much, because there are far more plausible ethical theories to be found in philosophical ethics. Thus, the second of our methodological lessons blocks us from interpreting the work of the researchers in ADJDM that we have examined so far as attempts to produce novel ethical theories.

Metaethical Neutrality

Our final methodological principle says that research in moral psychology should remain metaethically neutral, unless controls are included that reflect the possible truth of all philosophically plausible theories in metaethics. But since that is a *very* demanding standard, the practical implications of our third methodological principle are simply that data produced by experiments in moral psychology should not be used as evidence for any nontrivial metaethical conclusions. And of course, there are many researchers in moral psychology whose work is compatible with this principle. For example, it is hard to find researchers in developmental moral psychology who are willing to draw deep metaethical conclusions from their data, even when the developmental data produced using protocols that mirror the protocols of ADJDM. (Fu et al. 2014 and Kagan 1990 provide an entrance into the relevant literature.)

However, it is unfortunately fairly common practice in ADJDM-style moral psychology to draw metaethical conclusions from studies that lack even basic controls for the metaethical theories that are currently independently plausible in contemporary philosophy. In almost any other scientific discipline, these conclusions would be treated as nothing more than wild speculations; there is a sociological puzzle, then, about how adult moral psychology evolved a set methodological norms that are much less rigorous than those of most other fields in the behavioral and social sciences. Indeed, given the moral importance of moral psychology, and especially given the risks inherent in the scientific study of moral cognition, someone could be forgiven for assuming that the field would naturally developing extremely demanding methodological standards.

But I digress. Let us now take a look at some examples of research that violates the principle of metaethical neutrality. Our first example comes from the work of Fiery Cushman and Joshua Greene, who believe that a metaethical position that is a combination of intuitionism and noncognitivism can be derived from neuroscientific research (Cushman and Greene 2012, 277). They argue that their metaethical conclusion position should be accepted because it can explain some of the features of contemporary moral philosophy. They write:

In philosophy a debate can live forever. Nowhere is this more evident than in ethics, a field that is fueled by apparently intractable dilemmas. To promote the wellbeing of many, may we sacrifice the rights of a few? If our actions are predetermined, can we be held responsible for them? Should people be judged on their intentions alone, or also by the consequences of their behavior? Is failing to prevent someone's death as blameworthy as actively causing it? For generations, questions like these have provoked passionate arguments and counterarguments, but few clear answers.

Here, we offer a psychological account of why philosophical dilemmas [in ethics] arise, why they resist resolution, and why scientists should pay attention to them. ... We argue that dilemmas result from conflict between dissociable psychological processes. When two such processes yield different answers to the same question, that question becomes a "dilemma." No matter which answer you choose, part of you walks away dissatisfied. (Cushman and Greene 2012, 1)

So, philosophical ethics is not about, for instance, seeking out norms that when realized ameliorate poverty or injustice, and neither is philosophical ethics about discovering concepts and vocabularies that can make visible the situations of communities of people who are marginalized or disadvantaged by the practices of their society. Some professional philosophers do just that. Yet, incentives and endowment effects matter, and it is comparatively easier, given the popularity of the method, for a moral philosopher to advance her professional career by publishing papers that put forward increasingly refined definitions of a subtle yet precise conceptual distinction that communicates nothing more than the content of her intuitive judgments about a hypothetical moral dilemma. So, Cushman and Greene are not wrong to think that one (perhaps *the*) standard practice in contemporary Anglo-American philosophical ethics is to publish papers that are exceeding careful studies of the author's own moral intuitions, or if not that, very careful reflections on certain normative concepts of particular interest to the author. Certainly, there is no expectation that a top-tier publication in mainstream philosophical ethics must cite, or make use of, intellectual resources that are either historical or scientific in nature. But we can grant that Cushman and Greene have insight into certain intellectual by-product of the economic structure of professional philosophy without being lead to their metaethical conclusions. For example, they write, "because philosophical debate erupts at the fault lines between psychological processes, it can reveal the hidden tectonics of the mind" (Cushman and Greene 2012, 2). However, if the economic structure of professional philosophy changed only very slightly, so that it became, for instance, easier to get tenure by studying the social and intellectual history of gender constructs than by producing exquisitely refined analyses of increasingly recherché moral intuitions, then the practice of philosophical ethics would not be so amenable to explanation by Cushman and Greene's neurological thesis. The same point more plainly put: we should avoid attributing distinctive neurological causes to patterns of social behavior that can be explained by independent and fairly elementary economic analysis, and that point holds whether or not the patterns of behavior are rational.

That said, the problem with Cushman and Greene's actual argument for their intuitionist noncognitivism is very simple: they have produced no empirical data that are inconsistent with either cognitivism or naturalism or constructivism about either philosophical ethics or the ethical inquiry that can occur in any number of different institutional contexts. In the review article that I have been quoting so far in this section, Cushman and Greene summarize over a decade's worth of neuroscientific research that they believe supports their metaethical view. By my count, they review or cite twenty-two distinct studies that each provide evidence for a different neurological hypothesis—like, for example, the hypothesis that individuals with damage to the ventromedial prefrontal cortex are far more likely than healthy individuals to endorse harmful behavior in order to promote a greater good (Koenigs et al. 2007; see also Ciaramelli et al. 2007; Mendez, Anderson, and Shapira 2005). But none of these hypotheses are inconsistent with, for instance, cognitivism; none of the neuroscientific findings are even just simply hard to explain assuming independently plausible cognitivist premises. The same reasoning holds, mutatis mutandis, for the other metaethical positions—and so it is simply a mistake to draw any metaethical conclusions whatsoever from their neurophysiological data.

In fact, Cushman and Greene's literature review, rather ironically, helps make this conclusion more salient. Their review demonstrates that it takes a significant amount of cooperation, collaboration, networking, trust, and resource sharing to produce neuroscientific data (cf. Latour 1999). Yet none of the neuroscientific data reported by Cushman and Greene is incompatible with the idea that similar social practices may exist and sustain either moral practice or ethical inquiry. None of the data speak to the question of whether or not there is a community of philosophically sophisticated researchers who have built real patterns of cooperation, collaboration, networking, sharing, deliberation, and debate in order to understand and explain human normative behavior—and perhaps also guide it toward its own internally rational ends. True, that community might not be coextensive with the community of high-status moral philosophers employed at the leading research universities in England and America—but as we have learned from feminist epistemology of science, we should never assume that socially certified expertise aligns either perfectly, or even at all, with actual existing expertise (Anderson 2000). It is not impossible that the relevant community of philosophically sophisticated researchers who are making ethical progress are not themselves professional philosophers, and it is equally not impossible that such a community generates this progress by employing methods that place much less weight on the evidential import of moral intuitions.

For our second violation of the principle of metaethical neutrality, we turn to Jonathan Haidt's well-known social intuitionist model (Haidt and Joseph 2004; Haidt 2001). According to this model, moral cognition is deeply constrained by a set of about five modules, the function of which is to produce affective responses to five different classes of stimuli. When these modules produce negative affect, moreover, the mind is led to judge the stimulus as "wrong" or "bad"; likewise, when positive affect is created by the stimulus's interaction with the module, it is judged to be "good." According to Haidt and his collaborators, these moral-module-mediated affective responses are moral intuitions. What is added to this basic psychological story are a number of substantial claims about the subsequent impossibility of rational moral inquiry. The function of moral reasoning is not to articulate justification or explanations for moral claims; rather, it serves to stimulate further moral intuitions. Most of the time, reasoning provides nothing more than ad hoc justification for previously made and largely unchanging judgments. Because of this, moral reasoning and explanation do not truly justify judgments. And it is also very hard, if not impossible, for moral reasoning to escape the influence of the underlying moral intuitions; and these moral intuitions do not track whatever moral facts there may be. They are, instead, designed to produce affective responses that are largely in service of a number of evolutionary needs (Haidt 2001; Haidt and Bjorklund 2007).

The evidence that Haidt and his collaborators have failed to abide by the metaethical neutrality principle is the same as for Cushman and Greene discussed earlier. Haidt and his colleagues rely upon various different scientific studies to ground conclusions about the general impossibility of rational ethical inquiry (Haidt 2007, 2008)—and yet none of the relevant studies used protocols or controls that reflect the possibility that either cognitivism or naturalism or constructivism may be true. But rather than going through the whole body of evidence that has been produced over the last fifteen years for the social intuitionist model (SIM), allow me instead to provide a representative example of a failure to include controls that reflect a commitment to metaethical neutrality. The following paragraphs describe one of the famous, dumbfounding studies first run by Haidt and Scott Murphy, an undergraduate honors student in his lab.

> In study 1 we gave subjects 5 tasks: Kohlberg's Heinz dilemma (should Heinz steal a drug to save his wife's life?), which is known to elicit moral reasoning; two harmless taboo violations (consensual adult sibling incest, and harmless cannibalism of an unclaimed corpse in a pathology lab), and two behavioral tasks that were designed to elicit strong gut feelings: a request to sip a glass of apple juice into which a sterilized dead cockroach had just been dipped, and a request to sign a piece of paper that purported to sell the subject's soul to the experimenter for $2 (the form explicitly said that it was not a binding contract, and the subject was told she could rip up the form immediately after signing it). The experimenter presented each task and then played devil's advocate, arguing against anything the subject said. The key question was whether subjects would behave like (idealized) scientists, looking for the truth and using reasoning to reach their judgments, or whether they would behave like lawyers, committed from the start to one side and then searching only for evidence to support that side, as the SIM suggests.
>
> Results showed that on the Heinz dilemma people did seem to use some reasoning, and they were somewhat responsive to the counterarguments given by the experimenter. (Remember the social side of the SIM: people are responsive to reasoning from another person when they do not have a strong countervailing intuition). But responses to the two harmless taboo violations were more similar to responses on the two behavioral tasks: very quick judgment was followed by a search for supporting reasons only; when these reasons were stripped away by the experimenter, few subjects changed their minds, even though many confessed that they could not explain the reasons for their decisions. In study 2 we repeated the basic design while exposing half of the subjects to a cognitive load—an attention task that took up some of their conscious mental workspace—and found that this load increased the

level of moral dumbfounding without changing subjects' judgments or their level of persuadability. (Haidt and Bjorklund 2007, 197–198)

This is clearly not the right protocol to test the hypothesis that moral reasoning almost never aims a mind closer to the truth, or that it is impossible to make moral discoveries, or that it is impossible to implement some kind of procedurally fair decision protocol that can be used to decide the fundamental norms of a society, and so on. Indeed, these data could just as easily be interpreted as showing that participants trust their prior moral beliefs and explicit opinions more than they trust the experimenter who is arguing with them. What's more, note that Haidt and his collaborators put forward a comparative hypothesis—again: "The key question was whether subjects would behave like (idealized) scientists, ... or whether they would behave like lawyers"—but they collected no data about how scientists would reason about, for instance, unusual or weird causal problems outside of their area of expertise. But more on the problems with Haidt's dumbfounding study in chapter 6.

What is relevant here, instead, is that none of the data produced by Haidt is inconsistent with cognitivism, naturalism, constructivism, and so on. As we have previously noted, the kind of evidence that would confirm intuitionism must include both social and historical data, because it must falsify naturalism by demonstrating, for example, that a community set up to instantiate the patterns of collaboration that are characteristic of reliable inquiry in other domains has nevertheless failed, in the moral and ethical domains, to make progress toward realizing some of their most significant epistemic ends. Haidt and his fellow social intuitionists could compare the legal and political work done by the U.S. gay rights movement with the activities of the nascent Royal Society in developing late seventeenth-century chemistry (Shapin and Schaffer 1989), and determine whether the activities of one group are significantly less rational than the activities of another, or significantly less successful according to internal standards than the other. But for our purposes all that matters here is the fact that the failure to conduct such comparisons is evidence of a bias favoring of non-cognitivism on the part of Haidt and his collaborators, and which therefore provides us with another example of a failure to maintain metaethical neutrality within ADJDM-style moral psychology.

Conclusion

There are some specific ways in which philosophical ethics and moral psychology are coming apart. Part of the explanation of the inconsilience is ADJDM-style moral psychology's focus on only the most intuitive dimensions of moral cognition. The focus itself is unproblematic, since there are obviously important intuitive and nondeliberative aspects of moral cognition. But at the same time, it is important not to be mistaken about the methodological implications of the basic psychological fact that moral cognition is not entirely a matter of explicit, conscious, verbalizable, or socially mediated deliberation. There are moral intuitions, feelings, theories, and beliefs; and there are also moral habits, drives, plans, confusions, schemes, ambitions, ambivalences, and a host of other complex yet patently human psychological states and processes that are not so easy to subject to rational organization or assessment and yet are very much part of how our minds make up and interact with the moral world. Still, there is no inference from the existence of these psychological states and processes to any particularly deep conclusions about, for instance, the prospects of success or failure in philosophical ethics.

At a somewhat deeper level, however, it is instructive to ask why the intuitive elements of moral cognition have received the bulk of the professional attention of ADJDM-style moral psychology. As observed previously, the intuitive elements of moral cognition are relatively cheap to study—but that can be only a small part of the explanation. I suggest that a deeper explanation of current practice in ADJDM is that the discipline is responding to the problem that, technically, the best theories in philosophical ethics are unrealized, and therefore researchers in ADJDM-style moral psychology face a choice between developing theories that lack external validity (whether or not they are local or universal) and developing theories that lack some meaningful connection with philosophical ethics. My suggestion, then, is that the theories we have examined in this chapter are best interpreted as creative, even bold, attempts to develop a theoretical position that avoids either horn of that dilemma. By redefining core conceptual elements of philosophical ethics as psychological constructs, and by applying these constructs to only either neurophysiological processes, unrefined intuitions, or unreflective preferences; and by then investigating the question of whether an adequate explanation of moral thought and behavior can be developed on the basis of just these constructs, ADJDM-style moral psychology can be interpreted as seeking out a theoretical position that is both projectable from the view of philosophical ethics and that has strong prospects for universal validity. Unfortunately, this search has failed: seeking external validity by this route costs moral psychology its philosophical plausibility.

References

Adams, Robert Merrihew. 1999. *Finite and Infinite Goods: A Framework for Ethics. Oxford Scholarship Online*. New York: Oxford University Press.

Albee, Ernest. 2014 [1902]. *A History of English Utilitarianism*. New York: Routledge.

Anderson, Elizabeth. 2000. "Feminist Epistemology and Philosophy of Science." *The Stanford Encyclopedia of Philosophy*, August 9. http://plato.stanford.edu/entries/feminism-epistemology/ (accessed October 12, 2014).

Bartels, Daniel M., and David A. Pizarro. 2011. "The Mismeasure of Morals: Antisocial Personality Traits Predict Utilitarian Responses to Moral Dilemmas." *Cognition* 121 (1): 154–161.

Bleske-Rechek, April, Lyndsay A. Nelson, Jonathan P. Baker, Mark W. Remiker, and Sarah J. Brandt. 2010. "Evolution and the Trolley Problem: People Save Five over One Unless the One Is Young, Genetically Related, or a Romantic Partner." *Journal of Social, Evolutionary & Cultural Psychology* 4 (3): 115–127.

Boyd, Richard. 1991. "Observations, Explanatory Power, and Simplicity: Toward a Non-Humean Account." In *The Philosophy of Science*, ed. Richard Boyd, Philip Gasper, and J. D. Trout, 349–377. Cambridge, MA: MIT Press.

Boyd, Richard. 1990. "Realism, Approximate Truth, and Philosophical Method." In *Scientific Theories*, vol. 14, ed. Wade Savage, 355–391. Minneapolis: University of Minnesota Press.

Bruers, Stijn, and Johan Braeckman. 2014. "A Review and Systematization of the Trolley Problem." *Philosophia* 42 (2): 251–269. doi:10.1007/s11406-013-9507-5.

Christensen, J. F., and A. Gomila. 2012. "Moral Dilemmas in Cognitive Neuroscience of Moral Decision-Making: A Principled Review." *Neuroscience and Biobehavioral Reviews* 36 (4): 1249–1264.

Ciaramelli, Elisa, Michela Muccioli, Elisabetta Làdavas, and Giuseppe di Pellegrino. 2007. "Selective Deficit in Personal Moral Judgment Following Damage to Ventromedial Prefrontal Cortex." *Social Cognitive and Affective Neuroscience* 2 (2): 84–92.

Cichocka, Aleksandra, Michał Bilewicz, John T. Jost, Natasza Marrouch, and Marta Witkowska. 2016. "On the Grammar of Politics—or Why Conservatives Prefer Nouns." *Political Psychology*. doi:10.1111/pops.12327.

Connolly, Terry, and Lisa Ordóñez. 2003. "Judgment and Decision Making." *Handbook of Psychology, Part Three: The Work Environment* 19:493–517. doi: 10.1002/0471264385.wei1219.

Conway, Paul, and Bertram Gawronski. 2013. "Deontological and Utilitarian Inclinations in Moral Decision Making: A Process Dissociation Approach." *Journal of Personality and Social Psychology* 104 (2): 216–235.

Cushman, Fiery, and Joshua D. Greene. 2012. "Finding Faults: How Moral Dilemmas Illuminate Cognitive Structure." *Social Neuroscience* 7 (3): 269–279.

DeScioli, Peter, and Robert Kurzban. 2009. "Mysteries of Morality." *Cognition* 112 (2): 281–299.

Di Nucci, Ezio. 2013. "Self-Sacrifice and the Trolley Problem." *Philosophical Psychology* 26 (5): 662–672.

Ditto, Peter H., David A. Pizarro, and David Tannenbaum. 2009. "Motivated Moral Reasoning." In *Psychology of Learning and Motivation*, ed. Brian H. Ross, 307–338. Waltham MA: Academic Press.

Duke, Aaron A., and Laurent Bègue. 2015. "The Drunk Utilitarian: Blood Alcohol Concentration Predicts Utilitarian Responses in Moral Dilemmas." *Cognition* 134: 121–127.

Foot, Philippa. 2002. "The Problem of Abortion and the Doctrine of the Double Effect." In *Virtues and Vices*, ed. Philippa Foot, 19–32. Oxford: Oxford University Press.

Fu, Genyue, Wen S. Xiao, Melanie Killen, and Kang Lee. 2014. "Moral Judgment and Its Relation to Second-Order Theory of Mind." *Developmental Psychology* 50 (8): 2085–2092.

Gilovich, Thomas D., and Dale W. Griffin. 2010. "Judgment and Decision Making." In *Handbook of Social Psychology*, ed. Susan T. Fiske, Daniel T. Gilbert, and Gardner Lindzey, 542–588. New York: John Wiley & Sons, Inc.

Greene, J. D. 2010. "The Secret Joke of Kant's Soul." In *Moral Psychology: Historical and Contemporary Readings*, ed. Thomas Nadelhoffer, Eddy Nahmias, and Shaun Nichols, 359–371. New York: John Wiley & Sons.

Greene, Joshua D., Leigh E. Nystrom, Andrew D. Engell, John M. Darley, and Jonathan D. Cohen. 2004. "The Neural Bases of Cognitive Conflict and Control in Moral Judgment." *Neuron* 44 (2): 389–400.

Haidt, J. 2001. "The Emotional Dog and Its Rational Tail: A Social Intuitionist Approach to Moral Judgment." *Psychological Review* 108 (4): 814–834.

Haidt, J., and F. Bjorklund. 2007. "Social Intuitionists Answer Six Questions about Moral Psychology." In *Moral Psychology: The Cognitive Science of Morality*, vol. 2, ed. Walter Sinnott-Armstrong, 181–217. Cambridge, MA: MIT Press.

Haidt, Jonathan. 2007. "The New Synthesis in Moral Psychology." *Science* 316 (5827): 998–1002.

Haidt, Jonathan. 2008. "Morality." *Perspectives on Psychological Science: A Journal of the Association for Psychological Science* 3 (1): 65–72.

Haidt, Jonathan, and Craig Joseph. 2004. "Intuitive Ethics: How Innately Prepared Intuitions Generate Culturally Variable Virtues." *Daedalus* 133 (4): 55–66.

Hauser, Marc, Fiery Cushman, Liane Young, R. Kang-Xing Jin, and John Mikhail. 2007. "A Dissociation Between Moral Judgments and Justifications." *Mind & Language* 22 (1): 1–21.

Henrich, Joseph, Steven J. Heine, and Ara Norenzayan. 2010. "The Weirdest People in the World?" *Behavioral and Brain Sciences* 33 (2–3): 61–83, discussion 83–135.

Kagan, Jerome. 1990. *The Emergence of Morality in Young Children*. 1st ed. Chicago: University of Chicago Press.

Kirmayer, L. J., C. Fletcher, and L. Boothroyd. 1998. "Suicide among the Inuit of Canada." In *Suicide in Canada*, ed. A. Leenaars et al., 189–211. Toronto: University of Toronto Press.

Koenigs, Michael, Liane Young, Ralph Adolphs, Daniel Tranel, Fiery Cushman, Marc Hauser, and Antonio Damasio. 2007. "Damage to the Prefrontal Cortex Increases Utilitarian Moral Judgements." *Nature* 446 (7138): 908–911.

Körner, Anita, and Sabine Volk. 2014. "Concrete and Abstract Ways to Deontology: Cognitive Capacity Moderates Construal Level Effects on Moral Judgments." *Journal of Experimental Social Psychology* 55: 139–145.

Kurzban, Robert, Peter DeScioli, and Daniel Fein. 2012. "Hamilton vs. Kant: Pitting Adaptations for Altruism against Adaptations for Moral Judgment." *Evolution and Human Behavior: Official Journal of the Human Behavior and Evolution Society* 33 (4): 323–333.

Lanteri, Alessandro, Chiara Chelini, and Salvatore Rizzello. 2008. "An Experimental Investigation of Emotions and Reasoning in the Trolley Problem." *Journal of Business Ethics* 83 (4): 789–804.

Latour, Bruno. 1999. *Pandora's Hope: Essays on the Reality of Science Studies*. Cambridge, MA: Harvard University Press.

Lee, Jooa Julia, and Francesca Gino. 2015. "Poker-Faced Morality: Concealing Emotions Leads to Utilitarian Decision Making." *Organizational Behavior and Human Decision Processes* 126 (0): 49–64.

Mendez, Mario F., Eric Anderson, and Jill S. Shapira. 2005. "An Investigation of Moral Judgement in Frontotemporal Dementia." *Cognitive and Behavioral Neurology: Official Journal of the Society for Behavioral and Cognitive Neurology* 18 (4): 193–197.

Mikhail, John. 2011. "Appendix: Six Trolley Problem Experiments." In *Elements of Moral Cognition*, ed. John Mikhail, 319–360. Cambridge, UK: Cambridge University Press.

Navarrete, C. David, Melissa M. McDonald, Michael L. Mott, and Benjamin Asher. 2012. "Virtual Morality: Emotion and Action in a Simulated Three-Dimensional 'Trolley Problem.'" *Emotion* 12 (2): 364–370.

Rai, Tage S., and Alan Page Fiske. 2011. "Moral Psychology Is Relationship Regulation: Moral Motives for Unity, Hierarchy, Equality, and Proportionality." *Psychological Review* 118 (1): 57–75.

Rai, Tage S., and Keith J. Holyoak. 2010. "Moral Principles or Consumer Preferences? Alternative Framings of the Trolley Problem." *Cognitive Science* 34 (2): 311–321.

Royzman, Edward B., Justin F. Landy, and Robert F. Leeman. 2015. "Are Thoughtful People More Utilitarian? CRT as a Unique Predictor of Moral Minimalism in the Dilemmatic Context." *Cognitive Science* 39 (2): 325–352.

Scheffler, S., ed. 1988. *Consequentialism and Its Critics*. New York: Oxford University Press.

Shapin, S., and S. Schaffer. 1989. *Leviathan and the Air-Pump: Hobbes, Boyle, and the Experimental Life*. Including a translation by Simon Schaffer of Thomas Hobbes, Dialogus Physicus de Natura Aeris. Princeton, NJ: Princeton University Press.

Shweder, Richard. 2014. "Let Me Tell You a Story about Hindu Temples and Runaway Trolleys." In *Psychoanalysis, Culture, and Religion: Essays in Honour of Sudhir Kakar*, ed. D. Sharma, 270–282. New Delhi: Oxford University Press.

Swann, William B., Jr., Angel Gómez, John F. Dovidio, Sonia Hart, and Jolanda Jetten. 2010. "Dying and Killing for One's Group: Identity Fusion Moderates Responses to Intergroup Versions of the Trolley Problem." *Psychological Science* 21 (8): 1176–1183.

Thomson, Judith J. 1976. "Killing, Letting Die, and the Trolley Problem." *Monist* 59 (2): 204–217.

Uhlmann, Eric Luis, David A. Pizarro, David Tannenbaum, and Peter H. Ditto. 2009. "The Motivated Use of Moral Principles." *Judgment and Decision Making* 4 (6): 476–491.

Williams, Robert E., and Dan Caldwell. 2006. "Jus Post Bellum: Just War Theory and the Principles of Just Peace." *International Studies Perspectives* 7 (4): 309–320.

6 Moral Dumbfounding

Our next task is to examine a novel strategy for linking philosophical ethics and moral psychology; setting out the details of this strategy is the focus of most of the remaining chapters. However, before proceeding we must first examine the merits of a simple yet powerful objection to the line of argument that we have developed so far. While I have argued that some of the conceptual practices of ADJDM moral psychology lead to inconsilience between philosophical ethics and moral psychology, we have not yet considered the question of whether or not there *should* be inconsilience between philosophical ethics and moral psychology.

And as a matter of fact, the idea that ADJDM moral psychology should replace philosophical ethics has received a certain amount of discussion recently; removing philosophical ethics from the field of intellectual inquiry would of course render the question of consilience moot. Fiery Cushman and Joshua Greene put the idea this way:

> Where will [ADJDM] psychological research ... lead the field of philosophy? There is a perspective from which [this research] seems to undermine the foundation of philosophical debate. If rival [philosophical] claims simply reflect rival psychological systems, isn't the whole debate a charade? From this perspective, philosophers are just psychologists who take their conclusions too seriously, mistaking the psychology of the matter for the fact of the matter. (Cushman and Greene 2012, 277)

If there is psychological evidence that refutes, not cognitivism and naturalism as more general metaethical theories, but, specifically, the proposition that reasoning about ethical and moral properties typically can make a meaningful difference to the causes and content of moral judgment, then, since reasoning about ethical and moral properties is pretty much what philosophical ethics is, it would be a mistake to think that philosophical ethics can influence moral thought or behavior. Indeed, if the content of philosophical ethics could be explained away—and thus debunked—by the results of ADJDM moral psychology, that would suffice to demonstrate that philosophical ethics is pointless—and that moral philosophy may be nothing more than an academic appendix left over from a time before the rise of modern science.

So, why should anyone think that philosophical ethics cannot generate ethical knowledge, or otherwise help shape moral reasoning and judgment? This chapter examines the experiments that purport to demonstrate the reality of a psychological phenomenon called "moral dumbfounding," for it is the putative reality of moral dumbfounding that has led some philosophers and psychologists to conclude that reasoning about ethical and moral properties cannot make a meaningful difference to moral judgments. Yet, we shall see that it is a mistake to believe in moral dumbfounding.

Moral Dumbfounding?

According to Cushman, Young, and Greene, "moral dumbfounding occurs when individuals make moral judgments that they confidently regard as correct, but then cannot provide a general moral principle that accounts for their specific judgment" (Cushman, Young, and Greene 2010). Andrew Sneddon explains the phenomenon this way: "moral dumbfounding occurs when someone confidently pronounces a moral judgment, then finds that he has little or nothing to say in defense of it" (Sneddon 2007, 731–732). But how is this phenomenon supposed to demonstrate that reasoning about ethical and moral properties cannot make a meaningful difference to moral judgments? Susan Dwyer spells out the argument the following way: "moral dumbfounding would seem to constitute strong evidence against rationalism in moral psychology. If subjects really were reasoning, explicitly considering moral principles or not, we should expect that when asked to say why they judge thus-and-so, they will be able to articulate that chain of reasoning, or, at the very least, mention one or more of the principles to which they were appealing" (Dwyer 2009, 276). So, the objection arises from the fact that the moral dumbfounding is treated as evidence that moral judgments are produced by psychological systems that are encapsulated from deliberation. Here is Dwyer again: "The phenomenon of moral dumbfounding shows that it is highly improbable that ordinary moral judgments issue from a consciously accessible process of reasoning over explicit moral principles or rules" (Dwyer 2009, 294). The challenge, then, is that since the method of philosophical ethics is to create and evaluate explicit reasons for normative hypotheses, philosophical ethics cannot hope to gain any psychological traction, since the evidence that philosophers produce for different moral ideas will be unable to change anyone's mind.

My response to the objection is straightforward. An examination of the dumbfounding experiments shows that no substantial psychological conclusions about the causal structure of moral cognition follow from the extant data. And it is very important for me to stress that this is not because the experiments were subtly biased, or because there are purely conceptual problems with the relevant studies, and so on. Most experimental work in the behavioral sciences has one or two of those philosophical problems, and most of the time, such philosophical problems are not reasons to be skeptical about hypotheses based upon the relevant experimental data. However, the truth in the case of moral dumbfounding is that there is a very mundane reason why it would be a mistake to believe in moral dumbfounding: no one has figured out how to code for and thus measure the phenomenon, and consequently no one knows yet how to generate the kind of scientific data that would differentiate moral dumbfounding from other far less scientifically and philosophically interesting patterns of verbal explanation. After all, it is not hard to find people of any age, educational background, or ethnicity who have difficulty verbally defending otherwise confident judgments about all sorts of issues—from the fair value of a nonrefundable airline ticket, to the likelihood that McLaren will win a grand prix before 2020, to what is wrong with fascism, or why the Maxwell's demon thought experiment is not a refutation of the second law of thermodynamics. Indeed, most of us are just such people. It really should be no surprise that in the moral domain judgment and justification can come apart, since they come apart in just about every other domain as well. The challenge for proponents of the reality of moral dumbfounding, thus, is not to show that sometimes people are dumbfounded about moral problems, but, instead, to show that people are categorically worse at justifying their moral judgments then they are in other domains.

Coding is one of the pathways by which new constructs are introduced into the psychological literature. A new concept is defined partly in terms of a number of experimental variables and observable effects, and then, typically, taught to research assistants who are largely ignorant of the relevant experimental hypotheses and methods. They are then asked to apply the concept to the raw data generated by the experiment, and if they independently "code" (mostly) the same data points in (mostly) the same way, this validates the use of the construct to describe the relevant data. And then, more generally, psychologists who are investigating roughly the same psychological processes and mechanisms will be expected to use approximately the same constructs to code their data. The thesis of this chapter, then, is that moral dumbfounding *qua* psychological construct has not been sufficiently well operationalized for it to be used as a scientific coding construct. To be sure, the construct has a great deal of intuitive appeal to ADJDM moral psychologists, and it now has a fairly well-defined inferential role in the theory of moral psychology—but it is premature to believe that the phenomenon referred to by the construct has been observed and measured scientifically.

In fact, you do not have to take my word for it. It was Jonathan Haidt who introduced the concept of moral dumbfounding to contemporary moral psychology in an as-yet-unpublished paper that he coauthored with Frederick Bjorklund and Scott Murphy, called "Moral Dumbfounding: When Intuition Finds No Reason" (Haidt, Bjorklund, and Murphy 2000). This paper was not publicly available during the period of time—roughly from 2000 to 2010—during which Haidt and his collaborators were developing their now famous social intuitionist model of moral cognition. Evidence that moral dumbfounding is real and easy to generate played an extremely central role in the social intuitionist's arguments for their model, which is why the theory was rather severaly undermined when, in 2010, Haidt posted to his academic web page a copy of the "Moral Dumbfounding" paper with the following comments added as a preface: "This is a temporary report of one study, based closely on the honors thesis of Scott Murphy. ... It is so time consuming to code all the videos that even now, in 2010, I still haven't fully analyzed that study for publication. And in social psychology one cannot simply publish a description of an interesting phenomenon [namely, moral dumbfounding], which is what this report is." The point, of course, is that Haidt and his colleagues have not yet figured out how to reliably code for moral dumbfounding; a decade is enough time to design a reliable coding scheme for even the messiest verbal explanation data.

But what if Haidt and his team attempt to code the data sometime in the future? Two considerations are important. The first is that the success of Haidt's theoretical work over the last decade makes it extremely unlikely that the raw data could be coded in an unbiased fashion by either Haidt or any of his collaborators. It is customary for some recoding and reanalysis of raw data to occur during the peer review stage of publication in psychology; but it is odd, and perhaps more than a little unsettling, that a scientific theory has been presented to the academic community on the basis of data that, in addition to being unpublished, has not also been coded and is therefore unanalyzed. It would be far better, scientifically speaking, if the raw data were coded by scientists without a vested interest in the success of the social intuitionist model. Second—this is a point I remember Richard Boyd making as early as 2005 (cf. Boyd forthcoming)—it is well to recall that Haidt's dumbfounding study was designed to test a comparative hypothesis: do undergraduates at large American universities, when presented with some usual moral problems, reason more like idealized scientists or lawyers? Because of that, any of the coding categories that are developed to refer to possible cases of moral dumbfounding must also be able to refer to potential examples of dumbfounding in several others domains (law and science at least). Yet, until data about the putative dumbfoundability of lawyers and scientists is collected, it is premature to construct coding categories that apply only to the raw data produced by study participants who were asked to respond to moral dilemmas. Far better would be a process that explored different ways of coding the responses of participants who responded to moral, scientific, and legal dilemmas, as implementing such a process can reduce the likelihood that inevitable accidental differences between the response to the three types of dilemmas are reified into scientific evidence that people are categorically worse at either explaining or justifying their moral judgments than their non-moral judgments.

The Experiments

Let us turn now to a closer examination of two of the best-known moral dumbfounding experiments, the first of which is from Haidt and his fellow social intuitionists. One of the earliest descriptions of the unpublished study comes from a paper that Haidt coauthored with Matthew Hersh; recalling a quote in the previous chapter, they write:

> Haidt, Bjorklund, and Murphy ... dubbed this behavioral pattern "moral dumbfounding," which is defined as "the stubborn and puzzled maintenance of a moral judgment without supporting reasons" (p. 6). They brought participants into the laboratory, presented them with stories about harmless taboo violations (including consensual incest and harmless cannibalism of a body in a morgue), asked for an initial judgment, and then argued with them. On these stories, when compared to a standard moral reasoning dilemma (Heinz dilemma; Kohlberg 1969), participants showed a pattern of quick judgment, slow justification, frequently saying "I don't know," and frequently admitting that they could not find a reason to support their judgment. (Haidt and Hersh 2001, 194)

So, the Haidt-Murphy protocol used an interview to generate data, which Haidt describes the following way: "Scott brought thirty UVA students into the lab, one at a time, for an extended interview. He explained that his job was to challenge their reasoning, no matter what they said" (Haidt 2013, 36). Haidt and Murphy were looking for those participants who when presented with a disgusting choice would refuse to make the choice, and would continue to refuse to make the choice despite Murphy's attempts to convince them otherwise. Haidt reports that stories about consensual incest and harmless consumption of the flesh of a cadaver produced that pattern of response, but, for reasons that should be clear by now, he provides literally no quantitative analysis of how often this response pattern was produced in comparison to other response patterns.

Does the informal observation of this pattern by experimenters who have a professional interest in this pattern "being real" warrant the conclusion that reasoning does not influence moral judgment? It does not. There are many other psychologically plausible explanations of the same pattern, most of which are utterly mundane, and all of which license no deep conclusions about the etiology of moral judgment. For instance, undergraduates may be reluctant to admit to a peer that they have no problem performing a disgusting act, and they may also have a second-order reluctance to admit the first reluctance to an experimental interviewer, especially if the interviewer seems intent on playing devil's advocate with the participant. The serious methodological point is that the Haidt-Murphy protocol needs to be redesigned so as to control for these various alternative explanations before it can produce data that support the theory that moral judgment is causally isolated from the influence of deliberative reasoning. And all of those improvements assume that the most damning flaw can be overcome and that an operationally useful definition of moral dumbfounding can be developed and applied to the relevant data.

Let us turn now to the second experiment, which was designed by Marc Hauser's lab at Harvard, and which is called the "Moral Sense Test" (Hauser et al. 2007). For the Moral Sense Test, research participants were asked to visit a website where they were presented with pairs of moral dilemmas. Participants were asked to form a judgment about the moral permissibility of a choice presented by the dilemma, and were then given the opportunity to justify their judgments. What the experimental participants did not know when they took the Moral Sense Test was that some were given dilemma pairs that presented the option of resolving the dilemmas by making judgments consistent with moral principles such as "it is less permissible to cause harm as an intended means to an end than as a foreseen consequence of an end" (Hauser et al. 2007, 15). The guiding hypothesis seems to have been that, if participants make intuitive moral judgments that are consistent with such moral principles, this means that knowledge of the principle is probably the cause of their judgment. Then, if participants, when asked to justify their judgments, either fail to articulate the relevant moral principle or fail to supply an alternative credible justification, this is taken to be conclusive evidence that the principle seemingly guiding moral intuitions is "unconscious and inaccessible" (Hauser, Young, and Cushman 2008, 134). The Moral Sense Test is designed, therefore, to show that experimental participants have unconscious knowledge that cannot be articulated in the experimental setting.

Hauser and colleagues (Hauser et al. 2007; Hauser, Young, and Cushman 2008) did find that many of their participants had intuitions that were consistent with moral principles like the principle of double effect. Hauser and his collaborators also found that, when asked to justify their judgments, most of their participants not only failed to articulate moral principles that were consistent with their judgments, but were routinely unable to provide what, according to Hauser and his team's standards, counted as credible justifications for their moral judgments. So, in short, Hauser and his team found that people in the course of a fairly short experiment are routinely unable to adequately justify moral judgments that are seemingly quite sensitive to even quite abstract moral properties. They then took this result to warrant the conclusion that the moral judgments in question "are ... mediated by unconscious and inaccessible [moral] principles" (Hauser, Young, and Cushman 2008, 134).

As before, we can ask: do these data warrant any deep conclusions about the etiology of moral judgment? They do not, and the evidence for this negative answer rests in a clear understanding of the standard of justification that was used for coding responses to the Moral Sense Test. The justifications provided by participants in the Moral Sense Test were coded as either *sufficient justification, insufficient justification,* or *discounted justification.* A sufficient justification was one that identified any explicit factual difference between the two scenarios and referred to the difference as the basis of moral judgment. The complaint here is that the latter two categories were operationally defined so that they either imposed extremely demanding standards on participants, or assumed philosophically implausible standards of justification. For an example, Hauser and his team (Hauser et al. 2007; Hauser, Young, and Cushman 2008) defined the category insufficient justification to include justifications that combined utilitarian and deontological reasoning. But as we have seen, the constructs "deontological reasoning" and "utilitarian reasoning" have, as they are used in ADJDM moral psychology, nothing to do with philosophically plausible versions of deontology or utilitarianism. And what's more, some of the most sophisticated theories in philosophical ethics synthesize utilitarian, Kantian, and constructivist approaches to ethics; philosophers will appreciate the irony of the fact that, according to Hauser et al.'s protocol, Derek Parfit would be found to have produced insufficient justification for his ethical theory (Parfit 2011a, b). There really is no reason to think that combining these two "styles" of justification is evidence of some kind of fundamental division in the causal architecture of moral cognition.

More worrisome is how the category *discounted justification* was defined. "Responses that either were blank or in any way included added assumptions were discounted. Examples of assumptions include: (1) men walking along the tracks are reckless, while men working on the track are responsible, (2) the conductor is responsible for acting, while the bystander is not, (3) a man's body cannot stop a train, (4) the five men will be able to hear the train approaching and escape in time, and (5) a third option for action such as self-sacrifice exists and should be considered" (Hauser et al. 2007, 14). Note that (3) of course is a factual property of the vignette, and that (1) and (2) are alternative moral justifications. Quite clearly, the team behind the Moral Sense Test has not yet figured out how to define adequate versus inadequate justification so that they are operationally consistent.

We can see now the difficulty in both dumbfounding studies is the same. Neither group of researchers has established even a minimally valid coding scheme that can be used to determine just how bad the explanations are that participants in such experiments are able to provide. The available raw data shows that people's moral explanations are not logically or philosophically impeccable, of course. But what is needed in order to ground the inference from qualitatively imperfect justifications to causal conclusions about the organization of the moral mind is, *at the very least*, a scientifically credible coding scheme that can show that moral justifications are reliably worse than nonmoral justifications. So, there is no argument to be made from the dumbfounding studies to the conclusion that inconsilience between moral psychology and philosophical ethics is warranted, because there is no scientific evidence that moral dumbfounding is a real psychological phenomenon (cf. Keil 2006; Mills and Keil 2004).

Setting the Record Straight

At this point, a brief digression is called for in order to set the historical record straight. Discussions of dumbfounding are often linked with rejections of Kohlberg's moral psychology. Here is Dwyer one last time: "moral dumbfounding falsifies Kohlbergian rationalism" (Dwyer 2009, 279). In more detail, Tania Lombrozo expresses the relevant argument the following way:

> Kohlberg, for example, studied explicit moral reasoning by eliciting justifications for moral judgments, which suggests such reasoning is the basis for judgments (Kohlberg 1969). However, more recent approaches to moral cognition challenge assumptions about the coherence of explicitly held moral commitments and about their role in generating moral judgments. For example, a phenomenon called "moral dumbfounding" reveals that many moral judgments are not accompanied by *adequate* justifications (Haidt 2001). This suggests that the moral commitments identified in justifications are not causally responsible for the corresponding judgments (Haidt 2001; Hauser et al. 2007, 4). (Lombrozo 2009, 274)

The truth of the matter, however, is that Kohlberg relied upon a protocol that was for all meaningful scientific purposes neutral with respect to whether or not the justifications that he and his colleagues elicited were *qualitatively* good. Kohlberg was uninterested in the "adequacy" of the justifications that participants in his studies provided; his interest was only in the conceptual content of people's explanation, and not also their quality. Here is a detailed description of Kohlberg's experimental protocol:

Three forms of moral judgment interview were constructed. Each form consists of three hypothetical moral dilemmas, and each dilemma is followed by 9–12 standardized probe questions designed to elicit justification, elaborations, and clarifications of the subject's moral judgments. For each dilemma these questions focus on the two moral issues that were chosen to represent the central value conflict in that dilemma. For example, the familiar "Heinz dilemma" ... is represented in Standard Scoring as a conflict between the value of preserving life and the value of upholding the law. ... Of course, the dilemma can also be seen to involve other value conflicts, for example, between the husband's love for his wife (affiliation) and the druggist's property rights (property). In order to standardize the set of issues scored for each interview, the central issues for each dilemma were predefined. This pre-identification of standard issues not only allows the sampling of moral judgments about the six issues (per interview) for each subject, it also allowed the construction of three parallel forms of the moral judgment interview. [...]

The first step in Standard Issue Scoring involves the classification of the subject's responses to a dilemma into the two standard issue categories. This is a fairly simple procedure. In the Heinz dilemma, for example, all responses arguing for stealing the drug are classified as upholding the life issue; all those arguing against stealing the drug are classified as upholding the law issue. In the follow-up dilemma (Should Heinz be punished if he does steal the drug?) all responses arguing for leniency are classified as morality/conscience; all those arguing for punishment are classified as punishment.

The first step in scoring responses to a dilemma, then, is to separate material into the two issue categories. Since the issue units are large and often contain a great deal of material, Standard Issue Scoring involves two further subdivisions [norm and element] before stage scoring begins. This results in a fairly small unit of analysis. The Standard Scoring unit, the criterion judgment, is defined by the intersection of dilemma X issue X norm X element. Classification of responses by issue involves determining which choice in the dilemma is being supported or which of the two conflicting issues is being upheld. Classification by norm is a further subdivision of the interview material by its value content. (Colby et al. 2011)

Just so it is perfectly clear, here is a (slightly reworded) list of the coding constructs that Kohlberg and his collaborators used to generate their analytical data. Note that these categories are neutral with respect to the question of the epistemic quality of the responses. Each is defined as a categorical variable, which means that coding the data required only determining whether a participant did or did not say things that are described by the relevant construct.

Obeying or consulting persons or deity; blaming or approving; retributing or exonerating; having a right or having no right; having a duty or having no duty; reputation (good or bad); seeking a reward or avoiding punishment; consequences for the individual faces the choice (good or bad); consequences for the group the individual faces a choice is part of (good or bad); upholds character or upholds self-respect; serves a social ideal or social harmony; serves human dignity or human autonomy; balances perspectives or attempts to see the action from the perspective of another; considers reciprocity and positive desert; maintains equality or procedural fairness; maintains social contract. (Colby et al. 1983, 9–10)

These constructs were then collected together into higher-level categories that Kohlberg and his collaborators used as evidence for their theory of moral development. But for our purposes, only two observations are relevant; we do not have to consider the question of whether Kohlberg's overarching theory is correct. First, and again, none of these constructs are defined in terms of the quality of the reasoning generated by participants. Second, and rather telling, notice that these categories could not be applied to verbal responses *if it were not also the case that, in Kohlberg's studies (of which there were very many), people did not routinely provide fairly rich and detailed explanations for their moral judgments*. It would be an empirical mistake, then, to take Haidt's unpublished account of how his participants behaved as characteristic of how people respond to hypothetical moral problems (for more, see Fedyk and Koslowski 2013). Indeed, Kohlberg's coding variables could easily be applied to either Hauser's or Haidt's raw data, and that is very nearly a proof that the moral dumbfounding studies have not falsified Kohlberg basic commitment to the empirical proposition that people can reason about their own moral judgments.

Conclusion

Why does it matter that we get to a correct understanding of Kohlberg's methodology? For many researchers involved with ADJDM moral psychology, the putative failure of Kohlberg's "rationalism" is an invitation to take seriously the idea that moral intuitions about hypothetical dilemmas are expressions of innate moral instincts. Jonathan Haidt says of his social intuitionist model of moral cognition, which states that nearly all moral judgments are really just cognitive echoes of a small number of fundamental moral intuitions, "the SIM is essentially E.O. Wilson's [sociobiological] theory of moral judgment but with more elaboration of the social nature of moral judgment" (Haidt 2008, 69). If it can be shown both that patterns in people's moral intuitions would, if acted upon, generate adaptive patterns of behavior, then it seems that moral psychology may well have discovered the innate moral sense anticipated by Wilson, Joyce, and others. If there really are moral instincts, and there really is not much more to moral cognition beyond the expression through intuition of these instincts, then the set of moral behaviors will be a substantial, if not total, subset of the set of adaptive behaviors. By that sequence of inferences, evidence of moral dumbfounding becomes evidence both of the irrelevance of philosophical ethics—and of the importance of building deeper connections between evolutionary psychology (which itself relies on predictive explanations; see Fedyk 2014) and moral psychology. However, it is important to keep in mind one of the deeper implications of the argument from chapter 2: evolutionary biology provides no reason to posit the existence of innate moral instincts. No fact in either the practice or the theory of contemporary evolutionary biology requires that what moral intuitions there are be conceptualized as the expression of developmentally-endogenous moral instincts.

Still, we are left with a very different problem that arises out of the metaphysical fact that the most plausible theories in philosophical ethics are, technically, unrealized. Said another way, there no naturally occurring populations whose social behavior realizes the regulative framework of either Millian utilitarianism or Kantian deontology, or any other historically important ethical theory in moral philosophy. (Saying this is not also to deny the very real influence that many of these ethical theories have had over the centuries.) As we have previously noted, this creates a methodological problem for moral psychology: how can researchers in moral psychology produce theories that both have a nontrivial degree of external validity and which are also projectable from the perspective of philosophical ethics? Chapters 7 and 8 chart a route past this problem.

For now, it is interesting to conclude our critique of moral dumbfounding by observing that moral dumbfounding experiments are not the only kind of evidence that would suggest that contemporary philosophical ethics is a charade. Indeed, the idea that philosophical ethics is a pointless, or worse, is hardly new: generations of Marxists have argued for the idea that the real function of philosophical ethics is nothing more than to ratify the interests of the powerful in capitalist society. Brian Leiter puts one of these arguments this way: "the British intuitionists [like W. D. Ross] set the stage for the paradigm of moral philosophy over the last century by limiting its domain to allegedly obvious 'intuitions' shared by members of the same socioeconomic class about matters that never had systemic import" (Leiter 2015, 11). Note that if Leiter *is* right, then the focus on moral intuitions in ADJDM moral psychology means that not only does moral psychology serve the same ideological ends as early twentieth-century moral philosophy, it also amplifies the likelihood that these ends will be reified in contemporary society. It is one thing for moral philosophers to propose that certain normative principles ought to be accepted for realization on the grounds that the same moral philosophers find the principles intuitive; it is quite another for a high-prestige branch of science to "discover" that the human mind cannot help but accept the very same principles, because evolution built the human mind so that it cannot help but be motivated by the relevant principles.

And suppose, furthermore, that the "discovered" normative principles just happen to align extremely well with the liberal political project in the West (this is not a purely hypothetical idea; see Mikhail 2011); how could a critic respond in a manner that would be generally persuasive, except by starting up an alternative but equally well-resourced research program in the behavioral sciences? Scientism is the vitiation of meaningful comparative, historical, and philosophical inquiry, the effect of which is often to profoundly increase the cost of genuinely pluralistic inquiry in the moral sciences. The risk of scientism in moral psychology, then, is that the cost of "scientific" or "credible" or "serious" or "rigorous" or "projectable" research reaches such a level that the only people who can afford the price are also the people whose lived situations, and those of their patrons, are uniquely well served by the relevant normative discoveries. That is one reason why it is likely impossible for contemporary moral psychologists to even hypothesize that intuitions favorable to, for instance, socialism with Chinese characteristics are really the intuitions that got built into the architecture of human cognition in the Pleistocene.

References

Boyd, R. Forthcoming. "On Being 'Oh So Scientific': Evolutionary Theory as Methodological Anesthesia in Moral Philosophy and Moral Psychology." In *Problems of Goodness: New Essays on Metaethics*, ed. B. Reichardt. New York: Routledge.

Colby, A., L. Kohlberg, A. Abrahami, J. Gibbs, and A. Higgins. 2011. *The Measurement of Moral Judgment.*. New York: Cambridge University Press.

Colby, Anne, Lawrence Kohlberg, John Gibbs, Marcus Lieberman, Kurt Fischer, and Herbert D. Saltzstein. 1983. "A Longitudinal Study of Moral Judgment." *Monographs of the Society for Research in Child Development* 48 (1/2): 1–124.

Cushman, Fiery, and Joshua D. Greene. 2012. "Finding Faults: How Moral Dilemmas Illuminate Cognitive Structure." *Social Neuroscience* 7 (3): 269–279.

Cushman, F., L. Young, and J. D. Greene. 2010. "Our Multi-System Moral Psychology: Towards a Consensus View." In *The Moral Psychology Handbook*, ed. J. Dories, 47–71. New York: Oxford University Press.

Dwyer, Susan. 2009. "Moral Dumbfounding and the Linguistic Analogy: Methodological Implications for the Study of Moral Judgment." *Mind & Language* 24 (3): 274–296.

Fedyk, Mark. 2014. "How (not) to Bring Psychology and Biology Together." *Philosophical Studies* 172 (4): 949–967.

Fedyk, Mark, and Barbara Koslowski. 2013. "Intuition Versus Reason: Strategies People Use to Think About Moral Problems." In *Proceedings of the 35th Annual Conference of the Cognitive Science Society*, ed. M. Knauff, M. Pauen, N. Sebanz, and I. Wachsmuth, 424–429. Austin, TX: Cognitive Science Society.

Haidt, Jonathan. 2001. "The Emotional Dog and Its Rational Tail: A Social Intuitionist Approach to Moral Judgment." *Psychological Review* 108 (4): 814–834.

Haidt, Jonathan. 2008. "Morality." *Perspectives on Psychological Science: A Journal of the Association for Psychological Science* 3 (1): 65–72.

Haidt, Jonathan. 2013. *The Righteous Mind: Why Good People Are Divided by Politics and Religion*. New York: Vintage Books.

Haidt, Jonathan, Fredrik Bjorklund, and Scott Murphy. 2000. "Moral Dumbfounding: When Intuition Finds No Reason." Unpublished manuscript, University of Virginia. https://view.officeapps.live.com/op/view.aspx?src=http://faculty.virginia.edu/haidtlab/articles/manuscripts/haidt.bjorklund.working-paper.when%20intuition%20finds%20no%20reason.pub603.doc (accessed February 10, 2015).

Haidt, Jonathan, and Matthew A. Hersh. 2001. "Sexual Morality: The Cultures and Emotions of Conservatives and Liberals." *Journal of Applied Social Psychology* 31 (1): 191–221.

Hauser, Marc, Fiery Cushman, Liane Young, R. Kang-Xing Jin, and John Mikhail. 2007. "A Dissociation Between Moral Judgments and Justifications." *Mind & Language* 22 (1): 1–21.

Hauser, M. D., Liane Young, and Fiery Cushman. 2008. "Reviving Rawls' Linguistic Analogy: Operative Principles and the Causal Structure of Moral Actions." In *Moral Psychology*, vol. 2, ed. W. Sinnott-Armstrong, 107–144. Cambridge, MA: MIT Press.

Keil, Frank C. 2006. "Explanation and Understanding." *Annual Review of Psychology* 57:227–254.

Kohlberg, L. 1969. *Stage and Sequence: The Cognitive-Developmental Approach to Socialization*. New York: Rand McNally.

Leiter, Brian. 2015. "Why Marxism Still Does Not Need Normative Theory." *Analyse Und Kritik*,. http://papers.ssrn.com/sol3/papers.cfm?abstract_id=2642058 (accessed December 22, 2015).

Lombrozo, Tania. 2009. "The Role of Moral Commitments in Moral Judgment." *Cognitive Science* 33 (2): 273–286.

Mikhail, J. 2011. *Elements of Moral Cognition: Rawls' Linguistic Analogy and the Cognitive Science of Moral and Legal Judgment*. New York: Cambridge University Press.

Mills, Candice M., and Frank C. Keil. 2004. "Knowing the Limits of One's Understanding: The Development of an Awareness of an Illusion of Explanatory Depth." *Journal of Experimental Child Psychology* 87 (1): 1–32.

Parfit, Derek. 2011a. *On What Matters*. Vol. 1. Oxford: Oxford University Press.

Parfit, Derek. 2011b. *On What Matters*. Vol. 2. Oxford: Oxford University Press.

Sneddon, Andrew. 2007. "A Social Model of Moral Dumbfounding: Implications for Studying Moral Reasoning and Moral Judgment." *Philosophical Psychology* 20 (6): 731–748.

7 The Causal Theory of Ethics

We observed in chapter 5 that ADJDM moral psychology faces a difficult methodological dilemma. Because they must take considerations of external validity seriously, researchers in ADJDM must work with a theory of ethical thought and behavior that refers to at least some real-life patterns of cognition and action. But also, to sustain a projectable connection with moral philosophy, ADJDM moral psychology must employ at least some of the most entrenched concepts in moral philosophy. However, we have seen evidence that it is hard to concurrently apply both these principles in scientific practice, as the effect of following these principles is the emergence of inconsilience between philosophical ethics and moral psychology. This chapter introduces an ethical theory—called the causal theory of ethics—that is designed to allow moral psychology to avoid this methodological dilemma, by, technically speaking, introducing a philosophically plausible theory of ethics that is also about realized regulative frameworks. Said another way, the causal theory of ethics provides a platform that moral psychology can use to simultaneously satisfy the demands of philosophical projectability and external validity.

And indeed, there are several criteria that the causal theory must itself satisfy in order for it to be a useful for linking philosophical ethics with moral psychology. To wit:

1. *Philosophically projectable*: the theory should make projectable use of entrenched concepts from philosophical ethics, which allows the theory to provide a principled answer to the question of which social behaviors are ethical behaviors.
2. *Focused on the structural content of regulative frameworks*: the theory should be primarily about the structural content of regulative frameworks.

A fortiori, the theory is not a psychological theory; it should not attempt to describe the psychological representations, physical presuppositions, or enabling knowledge associated with any (realized or unrealized) regulative frameworks that may be found to fall within the theory's scope.

3. *Socially realistic*: the theory should imply that at least one realized regulative framework is an (or could be *the*) ethical regulative framework.
4. *Useful to moral psychology:* the theory can be used to generate empirical information that can be used to formulate tests of the local external validity of hypotheses in moral psychology, and possibly also tests of universal external validity.
5. *Methodologically compatible with evolutionary biology:* the theory must, if accepted, generate no implications that would lead to violations of (the general formulation of) Mayr's lemma.
6. *Scientifically projectable:* in addition to its philosophical plausibility, the theory must also be scientifically projectable. The theory, that is to say, must be plausible both in light of contemporary cognitive science, and more generally, contemporary social science as well.

These are interdependent criteria. For example, the stipulation that the theory is primarily about the structural content of regulative frameworks is a way of ensuring it is methodologically compatible with evolutionary biology, since this renders the theory compatible with the idea that a different theory (or, what is more likely, a family of alternative conceptual vocabularies, each of which) can speak to (a dimension of) the proximate foundations of whichever regulative frameworks are found within the scope of the causal theory of ethics. More simply: criteria (2) and (5) are different sides of the same conceptual coin.

Before we dive into the details of the theory, I want to say a word about (2) and (6), starting with the latter. You may recall that in chapter 4 we asked whether Rossian intuitionism was compatible with the background assumptions of contemporary cognitive science, and we found that the answer there was no. It is therefore appropriate for moral psychologists to simply ignore (even though it is *possibly* true) Rossian intuitionism for the purpose of designing and conducting empirical research. The reason for this is as simple as it is deep. It is just not possible to test all of the scientific hypotheses that are logically consistent with even the largest pools of available empirical data; scientists therefore should (and mostly only do) investigate just those scientific hypotheses that are likely to be true conditional upon the assumed approximate truth of previously accepted scientific theories. If the causal theory of ethics is to be genuinely useful to moral psychology (meaning: not ignored on grounds that it lacks scientific plausibility), it too must pass a similar test of scientific plausibility—and since the causal theory is a theory about, at bottom, the normative structure of different human populations, the relevant background theories against which its scientific plausibility can be established must come from the social sciences. It matters, then, that the causal theory of ethics be formulated using some of the core conceptual resources from certain social sciences, for this is one of the surest ways of ensuring that the truth (if indeed it is close enough to the truth) of the causal theory of ethics is not independent of the evidence that has used to justify the relevant background social scientific research.

Let me turn now to an implication of (2). This criterion entails that the causal theory of ethics have a nonpsychological theory of value, if it is to have one at all. And as we shall see, a theory of value is central to the causal theory of ethics. But this nonpsychological theory of value itself needs to be properly foregrounded, because it is presented as a thesis supported (but not confirmed) by empirical evidence, and this empirical grounding can make it very easy to misunderstand the tenability of the thesis. First of all, the theory of value in the following is not a theory of *all* of the things (properties, outcomes, patterns of distribution and so on) that are of ethical or moral value, if indeed there are any such things at all. Rather, the theory of value says that there may be a sui generis complex social outcome that is valuable—and that this kind of social outcome is composed of both the realization of particular outcomes that themselves are typically perceived to be valuable by the people whose behavior helps cause the particular outcomes, and the causal processes running between these outcomes in virtue of which particular valuable outcomes tend to be mutually reinforcing. I call the kind of complex social outcome I am talking about *the good and valuable*—but at the same time, I do not intent for this construct is to be taken literally or at face value. While the other conceptual components of the causal theory of ethics can and should be interpreted along broadly realist lines (meaning: these concepts are intended to assert probable truths), the appropriate interpretation for uses of *the good and valuable* is that of tactical nominalism. This means that my claims about the good and valuable are intended only to introduce a technical concept that stands for a kind that only future empirical research can determine to be real or not. Said differently, the construct *might* correspond to some real kind at the level of human populations: it could be that, just like species are real kinds caused by the interaction of complex population-level processes that are partly self-sustaining, so too may the good and valuable be caused by (very different) complex population-level processes that also are partly self-sustaining. But at the same time, empirical evidence of the existence of different realizations of the good and valuable is thin and mostly speculative, and partly for that reason, there is no scientific or philosophical tradition in which the construct has proved inductively useful. The construct, then, can at present function only as a linguistic signpost—a conceptual hypothesis—that calls attention to certain social phenomena that merit further investigation. Of course, my view is that this investigation is likely to vindicate the conclusion that the good and valuable is a real structural component of some (and certainly not all) normatively integrated populations; from my perspective, that is why the nominalism is, well, *tactical*. But at the same time, the evidence needed to confirm the existence of the good and valuable does not exist in any of the literatures I have knowledge of.

Social Institutions and Outcomes

We will start by characterizing what ethical inquiry can be about, at a very abstract level. What follows are some well-known data collated by the GapMinder Foundation from various underlying data sets that have themselves been produced by a number of different nongovernmental organizations, United Nations agencies, and the Bretton Woods Institutions. The data must be taken with a very large grain of salt; but the overall story the data tells is clear enough. Figure 7.1 shows data for nearly every country in the year 1800. On the y axis is life expectancy. On the x axis is gross domestic product in purchasing power parity, divided by the country's population. Each circle represents a country, and the diameter of the circle represents the relative size of the population in that country. And below, in figure 7.2, are the same numbers for about two hundred years later.

Not every population has done equally well. Yet there is a strong upward and rightward trend across in all the data points, and this trend is itself evidence of both a very general increase in wealth and health, and also of a

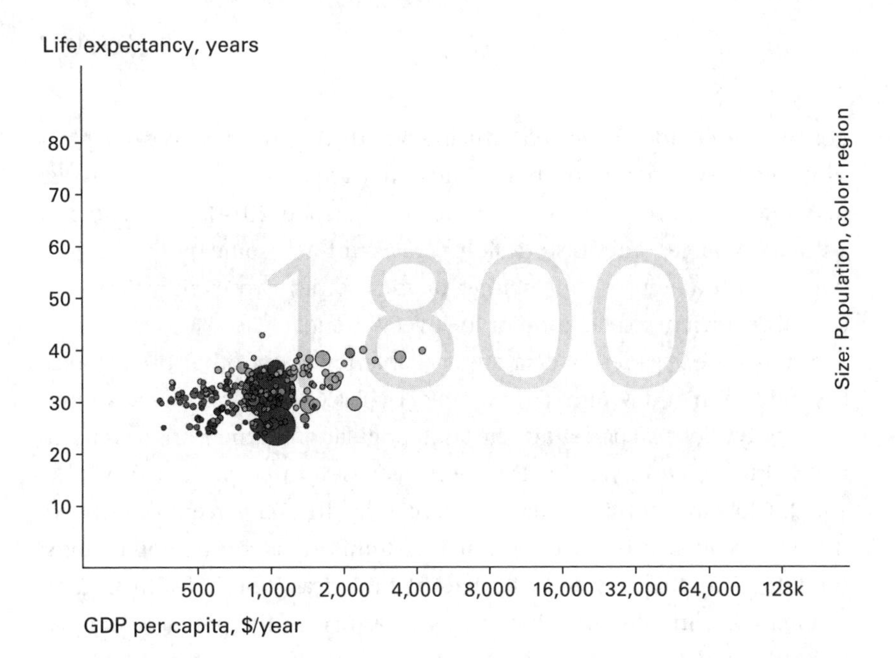

Figure 7.1
A rough representation of the health and wealth of the world's population, circa 1800.

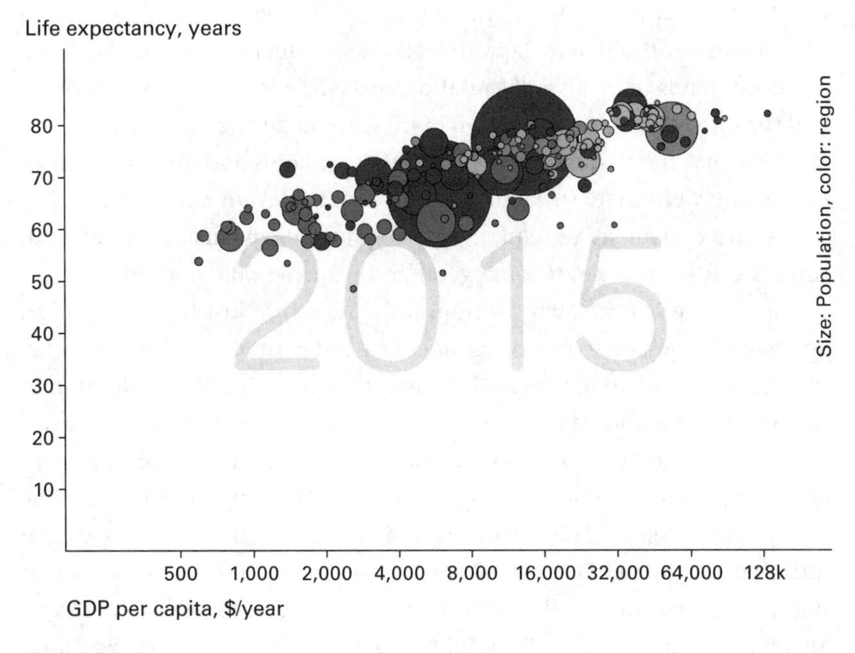

Figure 7.2
A rough representation of the health and wealth of the world's population, circa 2015.

positive correlation between health and wealth. And there is only one plausible causal explanation of these trends: they are caused by effective social institutions (Rodrik 2008; though see also Hassoun 2014). Differences in natural resources and historical luck account for some population-to-population differences, but neither of these other two explanations is a causally sufficient explanation of the overall trend.

Similar—very general—data are available for many other basic goods, beyond health and wealth. For example, UNESCO estimates that at around 1850, only about 10 percent of the adult population of the world was minimally literate, and by 2005, this level had risen to 80 percent (UNESCO 2005). Globally, mortality due to malaria has fallen 42 percent since 2000, though of course this rate of decline is not uniform across all populations; the rate of decline has been 49 percent in Africa (World Health Organization 2013). In 1860, child mortality measured as the probable number of deaths under the age of five per every thousand children in a country ranged from a low of approximately one hundred and seventy to a high of approximately five hundred. These figures remained roughly constant until the twentieth century, and they have now declined to lows of between five and twenty deaths per thousand and highs of around one hundred and fifty deaths per thousand (GapMinder 2014). Again, at a very general level, these are trends indicative of causation at the level of social populations—and the only plausible explanation of these trends at the same level of generality is that the trends are effects of different kinds and arrangements of (increasingly effective) social institutions in the relevant populations.

We arrive, then, at a preliminary metaphysical conclusion: social institutions can be a cause of some good and valuable outcomes in different populations. It is ultimately an empirical question of just how many good and valuable outcomes can be produced by different forms of institutional development and arrangement; likewise, it is also an empirical question to ask which good and valuable outcomes are such that they can be caused by some configuration of social institutions or another. In their most general form, these are questions that other scholars have pursued at some length (McCloskey 2010; Grootaert and Narayan 2004; Frey and Stutzer 2010; Acemoglu and Robinson 2013). However, our interest is not in that line of research; rather, the important point here is to observe a simple analytical distinction that is useful because it prevents us from overinterpreting the data shown earlier. For, and again, these data answer only a very basic metaphysical question: what *type* of thing is sometimes a cause of some good and valuable social outcomes? By abductive inference from the data, the answer is: social institutions. However, these data are of no use in answering a host of much more specific and difficult empirical questions—questions like, which particular good outcomes are caused by which particular institutions? Can we make any generalizations about types of institutions and their effect on types of good outcomes? So, we have a distinction between *easy* and *hard* questions that can be asked of the relationship between social institutions and the increasingly better (or worse) outcomes that can be observed in different populations.

Beyond that distinction, there are two more initial points that are useful to have in mind. The first is illustrated by a relationship we noted earlier, namely that, at a general level, wealth and longevity are positively associated in nearly every national population, and have been for roughly a century and a half at least. Still, the precise structure of this association cannot be read off such abstract data. Maybe increases in wealth cause increases in life expectancy, or maybe it is the other way around—or perhaps there is a range of different causal interactions between different realizations of wealth and different forms of longevity and a host of other mediating factors that are more or less unique to the different populations that, when viewed from a very abstract level, generate the data given earlier. The point for now is that there is a positive association of some kind and of some strength between wealth and longevity, and that this association is probably mediated by different social institutions in different populations; this is not also to say, baldly, that health causes wealth or vice versa.

The second point is about the metaphysics of social institutions. There are different theories about how social institutions are formed and what their common functions may be, if there are any such common functions. Yet what all of these accounts have in common is a commitment to the idea that, in order to realize a social institutions, certain clusters of norms have to be the basis of patterns of social behavior (North 1990; Searle 2010; Miller 2007). Whatever else they are, social institutions are realized clusters of norms, which means that social institutions are, by the definitions that we have introduced, *realized* regulative frameworks. Conceptually, then, some regulative framework must be realized by the behavior of a population of people in order to create a social institution, and social institutions therefore both construct and are caused by normatively integrated populations of people. Remember, according to the definitions introduced earlier in chapter 3, norms are inherently agent-neutral, which is why norms can be used to create social roles that can be occupied by different people over different periods of time—and which in turn implies that the social institution created by the realization of a regulative framework has an existence that is largely independent of the existence of any of the individual persons whose actions, at some particular time and in some particular place, contribute to its realization.

It is worthwhile to stress that these ideas are not just philosophically plausible, in light of (6) from the earlier list. Here is how Michael Tomasello elaborates some of the implications of this way of thinking about norms in the context of evolutionary anthropology:

> Like conventions, social norms operate not in second-personal mode but rather in agent-neutral, transpersonal, generic mode. First, and most basic, social norms are generic in that they imply an objective standard against which an individual's behavior is evaluated and judged. [...] Social norms emanate not from an individual's personal preferences and evaluations but, rather, from the group's agreed-upon evaluations for these kinds of things. Thus, when an individual enforces a social norm, she is doing so, in effect, as an emissary of the group as a whole—knowing that the group will back her up. Group-minded individuals thus enforce social norms because their collective commitment to a social norm means that they commit not only to following it themselves but also to seeing that others do, too[.] (Tomasello 2014, 88)

We have a number of basic observations in hand now. Social institutions are one type of thing that can causes some good and valuable outcomes at the level of social populations. Interventions on social institutions may therefore lead to comparatively better or worse outcomes, or they could make no difference. Technically speaking, social institutions are *realizations* of regulative frameworks—modifying the structural content of regulative frameworks can therefore be one way of amplifying or ameliorating valuable outcomes. Yet, the structural content of these regulative frameworks cannot be reduced to any particular pattern of individual thought or feeling. The norms that constitute the structural content of a regulative framework are, in a way, third-person descriptions of the behavior of a group over a period of time—realized norms are what a group both should be doing and is doing in some context. Furthermore, we have also seen that some good and valuable outcomes are, over long periods of time and in very large groups of people, positively correlated with another. At the very least, this shows that the realization of certain basic goods are compossible with one another.

Our next task is to develop these observations into a theory that meets the criteria listed earlier. The theory has fourteen theses that each extend, refine, or otherwise apply one or more of these more informal observations.

Fourteen Theses

We begin with a definition of the concept of an *ethical norm* that builds on our existing definition of a norm. Not all norms are, strictly speaking, ethical norms—so what makes the norms constituting the structural content of some particular regulative framework *ethical* norms? What makes some regulative framework an *ethical* regulative framework? Our first task is to provide an answer to this question that is both philosophically plausible and compatible with the idea that some ethical norms are realized.

A Conceptual Thesis

1. For some clusters of norms, what makes one of these clusters of norms a cluster of *ethical* norms is that the norms in the cluster either are, or should be, fundamental in any normative context involving their realization.

The idea of *fundamental in a normative context* needs to be explained. It is one of those ideas that, even though it cannot be defined logically, is nevertheless easy to find examples of. To create a normative context is simply to realize certain norms: norms create their own normative contexts in virtue of their realization. Normative contexts, thus, are simply where, when, and with whom we engage in normative self-governance. But these normative contexts are not exactly the same thing, metaphysically speaking, as realizing a specific regulative framework—because it is possible for several different regulative frameworks to be realized in one particular group at some particular time and at some particular place. Consider the regulative framework of Roman Catholicism and the regulative framework of hockey. These two regulative frameworks can be very easily mutually realized, and so a game of hockey can be a *normative context* in which both of these regulative frameworks are realized—if the players are mostly Catholic, that is. Now, what it is for a norm to be fundamental relative to a normative context is for it to be a norm that has priority over all other norms in that context. For instance, the norm "try to score a goal" is more fundamental than the norm "don't bump into people" in the normative context formed by a game of hockey. But for our examples, the Catholic norms that express in a prohibition on murder, among other things, are more fundamental than the norms of hockey. If a situation should arise in which a person faces a choice between following the norms of hockey and following Catholic norms, following the later norms would be what it means for the Catholic norms to be more fundamental than the norms of hockey. In the latter case, the normative context switches to the much broader context of living as a Catholic.

The proposal, then, is that an *ethical* norm either is, or should be, fundamental in a normative context, which is almost the same as saying that whenever it is possible to realize an ethical norm, a person should do so. What's more, this conception of ethical norms entails that ethical norms are fundamental only in relation to whichever other norms could also be realized in the same context. So, the first thesis of the causal theory of ethics does *not* say that an ethical norm is such only if it is fundamental in *all* normative contexts; there may be contexts in which it is not possible to realize *that* particular norm that is ethical in other contexts. Instead, the thesis says only that ethical norms have priority over nonethical norms in relevant normative contexts.

It is important that we find evidence that our first thesis is not a radical departure from how moral philosophers have traditionally thought about ethical norms. Technically, we must show that the definition given earlier is a projectable extension of conceptual practice in the history of moral philosophy. So, note that thesis 1 is a plausible reformulation of what, for example, R. M. Hare is speaking of when he offers *overridingness* as part of the list of criteria that distinguish the moral from the nonmoral (Hare 1981); or what Kurt Baier has in mind when he asks whether "moral reasons always defeat at least all self-interested and perhaps all self-anchored practical reasons" (Baier 1995, 197); or what Philippa Foot is referring to when she says that moral judgments are presumed by modern moral philosophy to bear "a special dignity and necessity" (Foot 1972, 308); or the concepts that Kohlberg is reasoning from when he predicts that the higher developmental levels of moral cognition will "embody a 'Kantian' sense of categorical oughtness ... [and] a sense of the 'moral law' as higher than the 'legal law' when the two are in conflict" (Wren, Edelstein, and

Nunner-Winkler 1990, 153). Thesis 1 is not, thus, a radical departure from how some moral philosophers have traditionally conceptualized ethical or moral norms or both.

We turn now to reintroducing some of the analytical ideas from the earliest chapters of the book.

Psychological Foundations of Social Institutions

2. The structural content of any regulative framework, realized or not, cannot be reduced to the psychological content, presuppositions, and enabling knowledge associated with the relevant regulative framework.
3. In order to realize a set of norms, people do not need to have exact psychological representations of the structural content of the relevant regulative framework.

These two theses restate some of the ideas that were developed in chapter 2. The simplest argument leading to their acceptance is that they can explain why it is not generally possible to derive interesting conclusions about the history and effects of specific social institutions from purely psychological premises (Wolff 1990). Furthermore, theses (2) and (3) are necessary components of the causal theory of ethics if the theory itself is to be compatible with Mayr's lemma, and so the theory needs to have a degree of methodological consilience with evolutionary biology.

Indeed, there is another inductive argument that provides support for the decision to include theses (2) and (3) in the theory. Given that a distinction between the population (or structural or ultimate) level of analysis and the individual (or proximate or psychological) level of analysis has been observed by most other theories about how selection acts on nonhuman species (Millstein 2006; Gilinsky 1986), it would be very surprising if the very same fundamental distinction did not also apply to efforts to describe the social behavior of humans. Nothing about natural selection implies that it is important to keep structural (ultimate) and proximate conceptual vocabularies separate, except that natural selection is a kind of structural process, and life is the by-product of complex interactions between structural and substructural processes. To understand this complex interaction, concepts that are apt for representing a structural process will need to be paired (in some overarching theoretical framework) with concepts that are apt for describing substructural process. Indeed, several centuries of research have provided powerful evidence of the metaphysical fact that the biological world is largely constituted by systems of systems of systems of systems (and so on), that partially overlap, block, inhibit, amplify, break, or sustain each other (Wimsatt 1994). The biological world is exceedingly complex. And so, on inductive grounds, it is just not likely that humans are the one species that has somehow freed itself from the complexity of biological reality, such that a single conceptual vocabulary will be apt for representing all that can be known about even just our social behavior. Instead, it is far more likely that, given our capacity for language and collective intentionality, our social behavior is even *more* complex than that of any other species—and so it is equally highly likely that even the best theories of our social behavior will need to enforce distinctions that have become de rigeur in evolutionary biology, like the conceptual distinction between structural (ultimate) and individual (proximate) explanations. Adopting theses (2) and (3) is therefore a way of creating methodological consilience between the causal theory of ethics and evolutionary biology, and thereby contributes to the scientific plausibility of the former.

The next five theses define the construct *the good and valuable*, but as explained previously, there is no argument here that the good and valuable truly exists. At most, this construct is a name that might refer to a real (and very complex) social outcome.

Metaphysics of the Good and Valuable

4. A realization of the good and valuable is a realization of a cluster of good or valuable outcomes for people in normatively integrated populations, where most of the components of this cluster of outcomes tend to causally aid, in or some other way stabilize, the realization of one or more of the other components of the cluster. Realizations of the good and valuable can have, thus, a form of homeostatic unity.
5. One of the additional factors that stabilizes a realization of the good and valuable is, within the relevant community, a sufficiently widespread conception (or set of roughly equivalent conceptions) of the good and valuable, the propositional content of which is at least a roughly accurate representation of some (and not necessarily all) of the component good or valuable outcomes that, together, form the relevant realization of the good and valuable.
6. This conception contributes more than just stability of the linguistic conventions necessary to create and sustain a conceptual vocabulary

that refers to components of a realization of the good and valuable. Because of the influence of this conception on motivation, planning, and preference, and because these factors in turn are part of what helps to cause a realization of the good and valuable, the conception's contribution to stabilizing the realization of the good and valuable can enter into feedback loops with the realization itself. But at the same time, realizations of the good and valuable are not defined by the content of any conception of the good and valuable.

7. There are different realizations of the good and valuable for both different historical and existing normatively integrated human populations. Said another way, if the good and valuable has been realized, then it has been realized in different forms that have a family resemblance with one another. There is therefore probably no paradigmatic form of, or instantiation of, the good and valuable.
8. Specific good and valuable outcomes also can be realized in various different ways.

The evidence for theses (4)–(8) is more empirical than philosophical, being abduction over a number of otherwise disparate observations. We have already seen that at a sufficiently general level, wealth and health tend to go together, and theses (4)–(8) assert that this is partly explained by the fact that people pursue both health and wealth. That both are *desired* by enough people is part of what *causes* them to go together. But that is not the end of the story of why these two outcomes are often positively associated. The conjecture here is that there are also properties of (different realizations of) health and wealth that, independent of people's motivation to acquire these outcomes, make it easier for these outcomes to occur together. Healthy people can work longer or study more, wealthy people can purchase more medical resources, and so on. So, what these five theses about the metaphysics of the good and valuable claim is that two different kinds of feedback loops or interactions typically will be realized by most good and valuable things: there are the feedback loops created as consequences of human motivational structure and behavior, and there are also feedback loops created by the causal relations that hold between specific good and valuable outcomes themselves. And so, one way of testing whether or not theses (4)–(8) describe a real kind of social outcome is to investigate whether or not models that describe the cause of one specific good outcome in some population must also include variables that refer to other valuable outcomes that have been realized in the same population—for the indispensability of these additional variables would be strong evidence of the causal homeostasis thesis (4) describes.

But what observational evidence is there that theses 4–8 *might* describe something real? We begin with an anecdote from an expert in social policy. Juan Somavia, former head of the International Labour Organization, describes the connection between conceptions of the good and valuable and realized good and valuable outcomes this way: "Women and men without jobs or livelihoods really don't care if their economies grow at 3, 5, or 10 percent per year if such growth leaves them behind and without protection. They do care whether their leaders and their societies promote policies to provide jobs and justice, bread and dignity, freedom to voice their needs, their hopes and their dreams and the space to forge practical solutions where they are not always squeezed" (Somavia 2011). It is not only that economic growth, jobs, justice, and bread are individually valuable. Rather, the point is that *all* of the valuable things Somavia names "go together" because they have the ability to reinforce each other: it is easier to have the space to make autonomous life decisions where there is also justice, dignity, and freedom of speech. These valuable outcomes are sometimes causally interdependent, which renders the interdependence itself something of value, insofar as it is worth preserving. Consequently, designing policies that balance good outcomes in an organic fashion is quite a different task, and much harder, than designing policies that aim to maximize a single component of the good and valuable, such as economic growth.

Of course, an expert anecdote cannot take us very far. That is why it is interesting to find that efforts are underway to construct reliable sociometric measures of the realization of particular valuable outcomes, and that the data produced by these projects supports the idea that there can be causally sustaining relations of interdependence among the relevant good and valuable outcomes.

Here are some of the details of the relevant projects. One motivation for designing new sociometric measures for valuable outcomes is *The Report by the Commission for the Measurement of Economic Performance and Social Progress* produced under the direction of Joseph Stiglitz, Jean-Paul Fitoussi, and Amartya Sen. The report is a response to the dearth of reliable statistical measures of the valuable outcomes that are not directly connected with advanced capitalist modes of economic production. The authors reach several important conclusions, and please take note of how they echo theses (4)–(8):

> What we measure affects what we do; and if our measurements are flawed, decisions may be distorted. Choices between promoting GDP and protecting the environment may be false choices, once environmental degradation is appropriately included in our measurement of economic performance. So too we often draw inferences about what are good policies by looking at what policies have promoted economic growth; but if our metrics of performance are flawed, so too may be the inferences that we draw. [...] What we measure shapes what we collectively strive to pursue—and what we pursue determines what we measure—the report and its implementation may have a significant impact on the way in which our societies looks at themselves and, therefore, on the way in which policies are designed, implemented and assessed. (Stiglitz, Sen, and Fitoussi 2010, 7)

A population that measures—or imagines, or conceptualizes, or believes in, and so on—too narrow a slice of the range of possible good and valuable outcomes may cut itself off from the possibility of realizing more of the relevant range. Said differently, it is impossible to realize some good outcome in a population if no one in the population can conceptualize it, and since statistical constructs developed by social scientists can be effective tools of public education (or ideology), a first step toward quite literally increasing the breadth and stability of different patterns of good and valuable outcomes can be to construct measures for the outcomes in question. Consequently, the report has inspired a number of research groups and organizations to explore different definitions of measures of what in this chapter we call realizations of the good and valuable. These efforts include both investigations into how and to what degree some valuable social outcome has been realized, as well as efforts to understand better the different conceptions of the good and valuable that are linked with the relevant patterns of realization—since after all, "the choice of relevant functionings and capabilities for any quality of life measure is a value judgment, rather than a [purely] technical exercise. [...] Measuring all these features requires both objective and subjective data" (Stiglitz, Sen, and Fitoussi 2010, 15).

Here are some examples of the relevant research. First of all, the OECD has begun work on a project called the *Better Life Index* (OECD 2013). The *Index* is best described for our purposes as a composite of two different types of data. The first type is data from many of the official statistics bureaus of mostly OECD countries, where the data describe the availability in the populations of these countries of certain more or less easy-to-measure valuable social outcomes—such as safety from crime, educational access and attainment, life satisfaction, and the availability of employment. These data provide some evidence of both the clustering and homeostatic unity described by thesis 4, because these data show that educational access and life satisfaction are positively associated across all of the countries reported.

The second type of data is even more interesting. It consists of aggregations of individual responses to survey questions that asked people to rank-order their preferences for valuable social outcomes represented in the first data set. So, the second pool of data represents different population-specific conceptions of the good and valuable. But also, as a technical matter, the survey data show that conceptions of good and valuable do correspond (though not perfectly) to the measures of realized good and valuable outcomes.

Taken together, both data sets show that people tend to have preferences that align reasonably well with the outcomes measured by the *Index*. And so the index as a whole can be thought of as providing guidance to would-be migrants: it provides a tool that can be used to learn about one's own preferences for different kinds of valuable social outcomes, and evidence about which countries produce patterns of outcomes that match best the expressed preferences.

The Gallup-Healthways Well-Being Index is a similar research program that attempts, in our terminology, to describe how the good and valuable is realized in a sample of fifty countries from around the globe, including China, India, the United States, and Nigeria. This index is formed out of five different composite constructs: the proprietary construct *purpose* and then several more standardized measures of social, financial, community, and physical well-being. These composite constructs are of course meant to refer to specific types of good social outcomes. More importantly, the report is premised on the idea that, of the elements in the well-being index, "each one has a direct effect on the others and can positively or negatively contribute to growth or decline in these areas" (Gallup and Healthways 2014, 14). The report goes on to describe some of the possible interactions: for example, "people with higher incomes are less frequently ill and less frequently have chronic diseases, which lead to income loss and poorer physical well-being," "people who feel good about their communities are more likely to be physically active and to interact more frequently with others," "individuals with high physical well-being can share healthy activities with those they love," and "physical well-being enables people to increase their financial well-being by avoiding unnecessary healthcare costs or extended or unpaid absences at work" (Gallup and Healthways 2014, 14–16). Again, the argument here is that theses (4)–(8) can explain why the Gallup-Healthways researchers are drawn to such conclusions.

Thus, the argument for a tactical nominalism about theses 4–8 goes like this. When researchers have set out to independently measure good and valuable outcomes, they have produced data that can be well explained on the assumption that theses 4–8 are at least approximately true. That of course does not provide scientific confirmation of the five theses—again, the data only indicate that the construct *the good and valuable* deserves to play a role in further investigation. However, the data does show something of more direct relevance to our overarching argument, namely that theses 4–8 have a degree of social-scientific plausibility and thus further strengthen the scientific projectability of the causal theory of ethics.

Our next cluster of theses returns us to the topic of social institutions.

Metaphysics of Social Institutions

9. Social institutions are *realized* regulative frameworks.
10. Social institutions are metaphysical individuals. Each social institution is unique and discrete, because each social institution has both its own sui generis structural profile of causal inputs and outputs, as well as its own unique internal structure defined by some regulative framework(s).

The first thesis about social institutions simply restates one of the ideas introduced earlier—that whatever else they are or can be, social institutions involve the realization of specific clusters of norms. And again, in my terminology that is the same as saying that social institutions are the realization of the structural content of some regulative framework(s).

There are a number of independent arguments for the tenth thesis. Philosophically, there is an argument from the metaphysical fact that social institutions have psychological foundations: since social institutions are partly complex by-products of (not only, but essentially) individual patterns of thought and behavior, there will be no way to duplicate a particular social institution without also duplicating many of the individuals out of whose actions the institution emerges. But that is just not possible. While one oxygen atom is just as good as another when forming different niacin molecules, one person will not be always fungible with another when trying to create a duplicate social institution. As Seumas Miller observes, "the nature of any institution at a given time will to some extent reflect the personal character of different role occupants, especially influential role occupants, e.g. the British Government during the Second World War reflected to some extent Winston Churchill's character" (Miller 2007).

A complementary argument for thesis (10) comes from developmental economics, where Eva Vivalt recently concluded her metaanalysis of impact evaluations (Vivalt 2014). Impact evaluations are measures of the effects of specific developmental interventions, when the intervention itself has been structured so that it produces information that can be used to answer the counterfactual question, What would the observed effects have been if the relevant intervention had not be performed? For example, if the developmental intervention aims to reduce poverty in a community by 10 percent over a five-year period, the intervention (that is, the new policy framework, or training program, or direct aid, and so on) must be designed and implemented in such a way that, if poverty does decline by 10 percent within five years, it will be possible to know if the observed decline was caused by the intervention. Developmental interventions that are designed to permit impact analyses are, therefore, designed to support causal inference, and so these studies are most often associated with efforts to conduct quasi-experimental field trials, or with interventions that can be randomized, and so on. Vivalt's analysis examined whether it is typically possible to generalize from the data produced by impact evaluations, and her results suggests that it is not. Her study examined "587 papers on 196 narrowly defined intervention-outcome combinations," and used a Bayesian model to assess the extent to which an intervention outcome of one type predicted that another intervention of the same type would yield the same outcome. Her results are mixed. She writes, "past results have significant but limited ability to predict other results on the same topic," because the degree of variability is quite high. Specifically, the likelihood that a successful causal intervention will generalize is slightly higher than 50 percent if the intervention is small and performed by an NGO, and slightly less than 50 percent if the intervention is large and performed by a governmental organization. Thus, even when interventions on existing institutional arrangements are designed to generate knowledge that the observed effects are caused by the relevant intervention, the generalizability of this knowledge to at least one other social context is limited—intuitively, not much better than a coin toss. And the metaphysical point that is important here is that Vivalt's findings can be explained by the hypothesis that social institutions are individuals. Two different social institutions will have different structures, and an intervention that modifies the structure of one social institution and thereby produces a particular effect will not be guaranteed to produce similar effects when applied to a different social institution.

Finally, we return to more purely conceptual matters: the final theses of the causal theory of ethics link our earlier definition of ethical norms with the metaphysics of social institutions to generate a method for conducting a form of ethical inquiry.

A Methodology for a Form of Ethical Inquiry

11. Knowledge of the causal effects of social institutions on the populations that realize them can be produced by the methods of different social science and humanities disciplines.
12. For norms referred to by 1: by default only, norms should be defined as ethical norms if there is a posteriori evidence indicating that the relevant cluster causes some good or valuable outcome.
13. For norms referred to by 1 or 12: "switching criteria" are the standards and evidence by which it is determined that a new cluster of norms should be fundamental in some normative contexts, and an existing set of norms that can be realized in those contexts should not be fundamental. The content of these switching criteria will be derived from understanding of the local structural processes and the local conception of the good and valuable, as well evidence about how well the good and valuable is realized for and in the relevant population. Empirical inquiry in these two areas will generate "switching criteria" by, for example, uncovering new norms that, if realized, are likely to be more efficient at causing the same good outcomes, or are able to cause additional goods without diminishing any other valuable outcome, or are able to diminish certain harmful outcomes, and so on. Switching criteria will almost always be evidence about how and by what means it is possible, in some particular normative context, to do *better*.
14. For norms referred to by 1 or 12 or 13: if in one historically delimited normative context a norm is found to be ethical, it does not follow a

> priori that the same norm will be ethical in any other independent normative contexts. That is because what makes a norm an *ethical* norm are its causal effects in a historical setting, and norms with exactly the same propositional content can nevertheless have different behavioral or structural effects in different normative contexts. Thus, if any of the norms that are determined by this theory to be *ethical* norms are also general or universal truths, it will only be because there exists a very large pool of a posteriori evidence that shows these "general ethical norms" produce the same good effects in almost any normative context whatsoever.

Theses (11)–(14) of the causal theory of ethics describe a research program that can be summarized as the attempt to discover which norms are responsible for causing existing good and valuable outcomes, which modifications and interventions on these institutions can lead to better outcomes, and where the one overarching goal is to make more fundamental and thus preserve those social practices that are found to lead to increasingly better outcomes. However, this research program is not an investigation into individual beliefs and feelings: it is the investigation of the possible and actual effects of variations in the structural content of different regulative frameworks.

Near the beginning of this chapter we encountered the distinction between easy and hard questions; here, it is important to see that a research program that implements the causal theory of ethics has the potential to answer hard questions. Such a program of research, that is to say, could discover which social institutions are causally responsible for producing specific good and valuable outcomes. And to be perfectly clear, the proposal here is not that this research program is somehow coextensive for all ethical inquiry. Rather, the proposal is that the causal theory of ethics describes only a form of ethical inquiry—a program of research that, again, can be used to determine, for at least some patterns of social behavior, whether or not they are ethical behaviors on philosophically principled grounds. That said, it is not usually the case that moral philosophers are open to distinguishing between different kinds of rational ethical inquiry; the most common methodological distinction is only between applied and theoretical ethics. But for that, many moral philosophers have simply assumed that there can be only one single conceptual and methodological framework for carrying out reliable, or at least rational, ethical inquiry. So it is again useful here to return to the idea of conceptual framework pluralism: my proposal is that the causal theory of ethics should not be seen as a competitor to more traditional forms of philosophical inquiry into human normativity, but, instead, that it be seen as a complementary theory that has, among its virtues, a deep connection with existing social realities.

In any case, allow me to review some of the arguments for theses (11)–(14). There is, first of all, a preponderance of academic research that substantiates thesis (11). This work can be divided, though only roughly, into quantitative research and qualitative historical research (Goertz and Mahoney 2012). Over the last half century, many of the social sciences have developed mathematical techniques that are able to assay the causal effects of different social institutions, or their parts, on human behavior—including path analysis, multilevel models, structural equation modeling, and directed acyclical graphs. A body of critical theoretical research about the application of different formal frameworks to social contexts has also developed in parallel to this research (Russo 2008; Freedman et al. 2010; Pearl 1995; Imbens and Rubin 1995; Fienberg, Glymour, and Spirtes 1995). Then, on the other side of the distinction, we find a very broad range of methodologies and disciplines that are, each in its own way, capable of describing and evaluating causal processes in the social world—including work in ethnography, cultural anthropology, and humanistic studies of ideology. And so, by the lights of the causal theory of ethics, research in any of these various areas may have already discovered that some cluster norms are ethical norms. And please keep in mind the underlying commitment to conceptual framework pluralism: the challenge here is not to assess which methodology is the single best way of studying social worlds. Nor is the critical task to analyze the theoretical output of these different methodologies with the goal of learning which single theory is most likely to be true. Instead, the much more difficult challenge is to develop a theoretical perspective that is able to place the different theories and methodologies in relation to one another, with the goal of organizing them together into a deeper yet more synoptic philosophical vision of human normative behavior.

The argument for thesis 12 is straightforward. It is an application of the metaphysical principle that a kind is formed by some collection of different individuals when the individuals typically are known to produce a common effect. For example, we correctly infer the presence of the kind "acid" when any number of different individual substances are observed to have the common effect of donating H+ ions to bases. Analogously, the data from the social sciences show that different social institutions sometimes cause good and valuable outcomes. Of course, these data do not tell us which social institutions (or which of their component regulative frameworks) are responsible for the beneficial outcomes. But all the same, the data do provide justification for applying the aforementioned principle. *Some* norms are causing these patterns of outcomes, and so there is license to treat these norms as sui generis—as, specifically, forming the kind *ethical norm*. (Or, technically, these data license the conclusion that one type of the kind *ethical norm* is the type of norm that causes good outcomes in a population; there are of course other categories of ethical norms that fall well outside of our present discussion.)

The argument for theses 13 and 14 is induction from the details of social scientific research on the connection between social institutions and valuable outcomes. No laws of human society have ever been discovered by social researchers, no matter which methodology they adopt or disciplinary framework they choose to work within. There is absolutely no reason, then, to expect that, when we use either social scientific or humanistic inquiry to examine the normative dimensions of different social worlds that are responsible for the production of valuable outcomes, either line of inquiry will eventually arrive at a small number of law-like generalizations about what configurations of regulative frameworks are conducive to such outcomes. A similar conclusion is nicely stated by Jon Elster: "the social sciences are light years away from the stage at which it will be possible to formulate general-law-like regularities about human behavior. Instead, we should concentrate on specifying small and medium-sized mechanisms for human action and interaction" (Elster 1989, viii). Nancy Cartwright and Jeremy Hardie have more recently reached a very similar conclusion from their analysis of public policy, and they recommend that we straightforwardly assume that it will not be possible to frame any law-like regularity about human behavior (Cartwright and Hardie 2012). So, theses 13 and 14 provide a way of translating these claims into the context of a form of normative inquiry. A categorical, or universal, or necessary ethical norm is, in effect, a law of human behavior; and while there may be such laws in a priori ethical theory, any ethical theory that posits laws will be implausible from the perspective of contemporary social sciences.

Accordingly, the causal theory of ethics provides a framework for rational ethical inquiry that does not require positing any laws of human behavior and thus create the risk of inconsilience with social science. Indeed, unlike many traditional ethical theories, the causal theory of ethics is designed to allow for empirical findings to guide the rational revision of which norms are, at some time and for some place, ethical norms. Evidence that a cluster of norms causes some particular good or valuable outcome is evidence that the norms are ethical. But evidence that some other cluster would do better would be, then, evidence that the other cluster should be fundamental, and so should be realized, and so should be regarded as *ethical* in the place of the first cluster of norms.

Conclusion

The causal theory of ethics satisfies the criteria that define an ethical theory that is philosophically plausible, scientifically plausible and, most importantly, useful to moral psychology for the guidance it can provide about what existing social behaviors are relevant to tests of external validity; but the theory obviously is not intended to provide a framework for all possible rational ethical and moral inquiry. Again, it is important to keep the underlying influence of conceptual framework pluralism in mind: the causal theory of ethics is offered as just one piece of the overall puzzle formed by an adequate theory of normative behavior. It is of course this book's thesis that moral psychology provides further pieces, but at the same time, it is well beyond the scope of this work to consider the many, many further pieces that may together yield a satisfactory picture.

And so, to recap, according to the causal theory of ethics, discovering that a particular cluster of norms are ethical norms will take a great deal of empirical research. Expertise and evidence about local customs and history will be as necessary for acquiring that knowledge as statistical and mathematical acumen. But the underlying point is that for all the difficulty in acquiring the evidence needed to learn that a cluster of norms is ethical, the causal theory of ethics ensures that the definition will refer to a set of existing social behaviors in virtually every normatively integrated population—those patterns of social behavior that cause good outcomes. Even more importantly, when the relevant social theory develops enough to describe the specific norms that are causally responsible for particular good or valuable outcomes, or goes even further to describe the interventions that would lead to better outcomes, then that social theory has managed to answer what we earlier called the "hard questions." This means that the social theory has described patterns of behavior that can, in turn, be among the subject matter of moral psychology.

But this is not yet enough to provide a bridge between moral psychology and philosophical ethics. It is one thing to show that, conceptually speaking, it is possible to ask and answer "hard questions"—but that is not enough to solve the problem of inconsilience between ADJDM moral psychology and philosophical ethics. What is needed, in addition, is evidence that existing research programs in the social sciences and humanities do routinely produce answers to hard questions, and therefore that there are existing bodies of scientific and philosophical evidence that can be mined by moral psychology for the best insights into which existing patterns of social behavior are ethical behavior and, consequently, which patterns of social behavior are pertinent to tests of local external validity.

References

Acemoglu, Daron, and James Robinson. 2013. *Why Nations Fail: The Origins of Power, Prosperity, and Poverty*. Reprint edition. New York: Crown Business.

Baier, K. 1995. *The Rational and the Moral Order: The Social Roots of Reason and Morality*. Chicago: Open Court Press.

Cartwright, Nancy, and Jeremy Hardie. 2012. *Evidence-Based Policy: A Practical Guide to Doing It Better*. Oxford: Oxford University Press.

Elster, J. 1989. *The Cement of Society: A Survey of Social Order*. Cambridge, UK: Cambridge University Press.

Fienberg, Stephen E., Clark Glymour, and Peter Spirtes. 1995. "Causal Diagrams for Empirical Research: Discussion of 'Causal Diagrams for Empirical Research' by J. Pearl." *Biometrika* 82 (4): 690–692.

Foot, Philippa. 1972. "Morality as a System of Hypothetical Imperatives." *The Philosophical Review* 81 (3): 305–316.

Freedman, D. A., D. Collier, J. S. Sekhon, and P. B. Stark. 2010. *Statistical Models and Causal Inference: A Dialogue with the Social Sciences*. Cambridge UK: Cambridge University Press.

Gallup and Healthways. 2014. "Gallup-Healthways State of Global Well-Being." http://info.healthways.com/hs-fs/hub/162029/file-1634508606-pdf/WBI2013/Gallup-Healthways_State_of_Global_Well-Being_vFINAL.pdf (accessed February 22, 2015).

GapMinder. 2014. "Under-Five Mortality Rate." http://www.gapminder.org/data/documentation/gd005/ (accessed February 22, 2015).

Gilinsky, Norman L. 1986. "Species Selection as a Causal Process." In *Evolutionary Biology*, ed. Max K. Hecht, Bruce Wallace, and Ghillean T. Prance, 249–273. New York: Plenum Press.

Goertz, G., and J. Mahoney. 2012. *A Tale of Two Cultures: Qualitative and Quantitative Research in the Social Sciences*. Princeton, NJ: Princeton University Press.

Grootaert, Christiaan, and Deepa Narayan. 2004. "Local Institutions, Poverty and Household Welfare in Bolivia." *World Development* 32 (7): 1179–1198.

Hare, R. M. 1981. *Moral Thinking: Its Method, Levels, and Point*. Oxford: Oxford University Press.

Hassoun, Nicole. 2014. "Institutional Theories and International Development." *Global Justice: Theory Practice Rhetoric* 7 12-27.

Imbens, Guido W., and Donald B. Rubin. 1995. "Discussion of Causal Diagrams for Empirical Research by J. Pearl." *Biometrika* 82 (4): 694–695.

McCloskey, D. N. 2010. *The Bourgeois Virtues: Ethics for an Age of Commerce*. Chicago: University of Chicago Press.

Miller, Seumas. 2007. "Social Institutions." *Stanford Encyclopedia of Philosophy*, January 4. http://plato.stanford.edu/entries/social-institutions/ (accessed March 3, 2015).

Millstein, Roberta L. 2006. "Natural Selection as a Population-Level Causal Process." *British Journal for the Philosophy of Science* 57 (4): 627–653.

North, D. C. 1990. *Institutions, Institutional Change and Economic Performance: Political Economy of Institutions and Decisions*. Cambridge, UK: Cambridge University Press.

OECD, ed. 2013. "Measuring the Sustainability of Well-Being over Time." In *How's Life? 2013: Measuring Well-Being*, 175–202. http://www.oecd.org/std/3013071e.pdf (accessed February 22, 2015).

Pearl, Judea. 1995. "Causal Diagrams for Empirical Research." *Biometrika* 82 (4): 669–688.

Rodrik, Dani. 2008. *One Economics, Many Recipes: Globalization, Institutions, and Economic Growth*. Princeton, NJ: Princeton University Press.

Russo, F. 2008. *Causality and Causal Modelling in the Social Sciences: Measuring Variations*. New York: Springer Science + Business.

Searle, John. 2010. *Making the Social World: The Structure of Human Civilization*. 1st ed. Oxford: Oxford University Press.

Somavia, Juan. 2011. "Time for a New Era of Social Justice Based on Decent Work." *World Day of Social Justice*. http://www.ilo.org/global/about-the-ilo/media-centre/statements-and-speeches/WCMS_151708/lang--en/index.htm (accessed February 17, 2015).

Stiglitz, J. E., A. Sen, and J. P. Fitoussi. 2010. *Report by the Commission on the Measurement of Economic Performance and Social Progress*. Paris: Institut National de la Statistique et des Études Économiques.

Tomasello, Michael. 2014. *A Natural History of Human Thinking*. Cambridge, MA: Harvard University Press.

Vivalt, Eva. 2014. "How Much Can We Generalize from Impact Evaluations?" http://evavivalt.com/wp-content/uploads/2014/10/Vivalt-JMP-10.28.14.pdf (accessed February 27, 2015).

Wimsatt, William C. 1994. "The Ontology of Complex Systems: Levels of Organization, Perspectives, and Causal Thickets." *Canadian Journal of Philosophy* 24 (sup1): 207–274.

Wolff, Robert Paul. 1990. "Methodological Individualism and Marx: Some Remarks on Jon Elster, Game Theory, and Other Things." *Canadian Journal of Philosophy* 20 (4): 469–486.

World Health Organization. 2013. "World Malaria Report 2013." http://www.who.int/iris/bitstream/10665/97008/1/9789241564694_eng.pdf. (accessed February 27, 2015).

Wren, T. E., W. Edelstein, and G. Nunner-Winkler. 1990. *The Moral Domain: Essays in the Ongoing Discussion Between Philosophy and the Social Sciences*. Cambridge, MA: MIT Press.

8 Hard-Question Social Theory

Is there evidence that hard questions get answered? That is, can moral psychologists find such answers in the output of the different academic fields that study the dimensions of human normative behavior? The main goal of this chapter is to provide examples of research that provide answers to hard questions and thereby count as identifying some of the social behaviors that are relevant to tests of external validity in experimental research in moral psychology. Of course, the examples that we will look at are not the only answers to hard questions that can be found in contemporary research. Indeed, because of the sheer breadth of research that can potentially answer hard questions, it cannot be our goal to survey this literature for the best examples of how researchers can go about providing answers to hard questions. The examples should therefore be taken as illustrations only.

One technical clarification before we proceed. When social research provides an answer to a hard question—to refresh: evidence that some particular good or valuable social outcome is, or could be, caused by a particular social intuition—the relevant social theory will have described at least one of only three conditions: either (1) there will be a worked-out proposal about how to modify an existing institutional arrangement (that is, revise the structural content of the regulative framework) so as to create a better outcome for some particular normatively integrated population; or (2) there will be an explanation of how an existing institutional arrangement produces a particular good or valuable outcome; (3) or there will be an explanation of how existing institutional arrangements produce a particular bad or detrimental outcome. Since only conditions 2 and 3 describe existing patterns of behavior, only hard-question social theories that describe those conditions are directly relevant to moral psychology. Still, it is easy enough to find examples of research in the social sciences and humanities that falls primarily within these two conditions.

From Institutions to Answers to Hard Questions

In fact, it is *very* easy to find such research that provides the relevant example. This ease itself is simple enough to explain, since after all the subject matter of hard-question social theory is phenomena that nearly all social scientists and humanists are interested in—namely, causal interactions that hold among social institutions, normatively integrated populations, and different kinds of valuable social outcomes. I have tried therefore to find examples that illustrate the diversity of different kinds of social theory that count as a form of ethical inquiry according to the causal theory of ethics. Given the causal theory of ethics, postcolonial sociology is as relevant to moral psychology as behavioral economics.

Interestingly, this implies that the kind of ethical inquiry framed by the causal theory of ethics is rather massively interdisciplinary, and because of that it will normally be impossible for any one person to become something like a general expert on ethical matters. And since these implications do not comport very well with the present state of professional philosophy—which has an incentive structure that presupposes that general ethical expertise can exist in an individual mind, thus making it practically irrational for someone who aims to be credentialed as a professional philosopher with an expertise in ethics to also spend their time and energy developing the kinds of interdisciplinary, collaborative networks that could begin to organize and thus render useful distributed ethical expertise—it may follow that professional philosophers are not the right kind of ethical experts to consult by the lights of the causal theory of ethics for answers to hard questions. Still, perhaps that only shows that the scope of what counts as philosophical ethics should be broadened so as to include more historically grounded and socially realistic intellectual traditions.

Five examples of hard-question social theory follow. Readers are invited to rely upon their own knowledge of research in the humanities and social sciences to see that there are many more possible examples of the same kind of research.

Happiness in the West

Robert Lane begins his book *The Loss of Happiness in Market Democracies* with the observation that "the economic and political institutions of our time are products of the utilitarian philosophy of happiness, but they seem to have guided us to a period of greater unhappiness" (Lane 2000, 13). Lane's research focuses on identifying the social institutions that cause happiness and unhappiness; according to his analysis, friendship and family are the strongest sources of happiness in many countries. People also tend to become happier as they become wealthier. However, once a market economy develops sufficiently to begin to destabilize the close relationships integral to family life and friendship, further growth in the market can cause people to become less happy. Lane's analysis is most directly applicable to the contemporary United States, and the various insights about the connection between social norms and outcomes in the United States are examples of answers to hard questions. For example, Lane finds a positive connection between happiness and the value society places on living with children (Lane 2000, 53); and more generally, Lane's results provide some prima facie support for deciding in Western democracies that we should treat norms that foster and protect friendship and family as more fundamental than norms that foster such things as, say, increasingly deep patterns of capital formation. Friends and family are good things, and happiness is too; the causal unity of these various good things is destabilized by baldly consumerist forms of life caused by excessive market growth.

Racism and Sexual Oppression in Anglo-America

Biopower is a term introduced by Michel Foucault for social practices that forcibly regulate and control the bodies and reproductive practices of different groups of people. It is fundamentally a causal notion, as it refers to the "more or less rationalized attempts to intervene upon the vital characteristics of human existence" (Rabinow and Rose 2006, 196–197). But unlike simple forms of Newtonian causation, the causal processes subsumed under the concept of biopower are extremely complex. Because they often involve the effects that widespread adoption of particular conceptual frameworks or ideologies impose on different groups of people, and because these effects often are realized over long periods of time and often create their own enduring feedback loops that, in turn, transform the relevant conceptual frameworks, the processes through which biopower is manifest are most often best studied using a historical and genealogical approach. In technical terms, it would be very hard to use a structural equation model to describe in full causal detail the different deleterious and beneficial effects produced by racializing different groups of people in the American South over the course of the nineteenth and twentieth centuries; an analytical framework organized around the concept of biopower does far better.

Ladelle McWhorter's *Racism and Sexual Oppression in Anglo-America: A Genealogy* is an example of just such an approach (McWhorter 2009). Her work provides a systematic history that details how racial and sexual minorities have been harmed by the various ways that governments and scientific authorities have both conceptualized and classified members of these groups. Her analysis exposes collective choices made by people in positions of power that effectively render patterns of oppression compatible with existing institutional frameworks. For example, on the basis of extremely detailed historical reports, she argues the following:

> Economic considerations played a major role in lynching, but economic calculation was not the sole driving force; white people were defending white power, white dominance, which by the late nineteenth century most whites held to be valuable in and of itself, far beyond financial measure. The point here is that when white people—especially newspaper editors, sheriffs, attorneys, and legislators—made claims that white fears of black male sexuality drove the lynching craze, they were stating something they knew to be less than entirely true. But they articulated and perpetuated this claim because, to the degree that it was believed, it enabled federal, state, and local law enforcement officials to characterize the phenomenon of lynching not as murder but as a sort of rogue juridical form, a ritual of popular justice. (McWhorter 2009, 158)

That of course is a causal conclusion, albeit one that depends upon a very subtle description of patterns of thought and behavior more or less proprietary to a very specific culture during a highly constrained period of time.

Nevertheless, by focusing on the causal processes tying together ideology and outcomes for various different groups of people that are revealed through applications of the concept of biopower, studies like McWhorter's provide a substantial amount of insight into the forces and barriers that different social groups face. Norms embedded in social institutions can block access to good and valuable outcomes just as easily as they can promote such outcomes, and when evidence of this blocking is produced, this is reason to eliminate or revise the relevant norms. A concrete example of this change would be replacing the cluster of norms according to which lynching is conceptualized by those who practice it as a cultural ritual with a cluster of norms for the same group of people according to which lynching is a deeply offensive crime. And it is of course a very different problem to figure out how just such a change might be undertaken. This is a difference to which will we return.

Ethics of Psychological Experimentation

Still, sometimes the causal effects of a cluster of norms are simple enough to be described in straightforwardly analytical terms, and Diana Baumrind's critique of an earlier version of the ethical code for the American Psychological Association (APA) provides a good illustration of this. Her critique is a response to Milgram's famous obedience study (Milgram 1963), and since Baumrind puts her critique more concisely than I can paraphrase it, permit me to quote Baumrind quoting Milgram:

> The detached, objective manner in which Milgram reports the emotional disturbance suffered by his subject contrasts sharply with his graphic account of that disturbance. Following are two other quotes describing the effect on his subjects of the experimental conditions:
>
>> I observed a mature and initially poised businessman enter the laboratory smiling and confident. Within 20 minutes he was reduced to a twitching, stuttering wreck, who was rapidly approaching a point of nervous collapse. He constantly pulled on his earlobe, and twisted his hands. At one point he pushed his fist into his forehead and muttered: "Oh God, let's stop it." And yet he continued to respond to every word of the experimenter, and obeyed to the end.
>>
>> In a large number of cases the degree of tension reached extremes that are rarely seen in sociopsychological laboratory studies. Subjects were observed to sweat, tremble, stutter, bite their lips, groan, and dig their fingernails into their flesh. These were characteristic rather than exception responses. ... On one occasion we observed a seizure so violently convulsive that it was necessary to call a halt to the experiment. (Baumrind 1964, 422)

After his experiment, Milgram did try to ensure that participants in his study left "in a state of well-being." Baumrind continues on to say, "It would be interesting to know what sort of procedures could dissipate the type of emotional disturbances just described. In view of the effect on subjects, traumatic to a degree which Milgram himself considers nearly unprecedented in sociopsychological experiments, his casual assurance that these tensions were dissipated before the subject left the laboratory is unconvincing" (Baumrind 1964, 422).

So, Baumrind and Milgram agree that participating in Milgram's obedience study caused a traumatic harm to many, if not most, of the participants. Milgram justifies this harm by appealing to the scientific gains, but our interest here is in Baumrind's analysis of the APA's *Ethical Principles in the Conduct of Research with Human Participants*. This document was written after Milgram's study, but what is not commonly reported when Milgram's research is talked about is that replications of Milgram's work were performed, including "with children as young as six" and that these replications include "graphic reports of the children's reaction [as consisting of] trembling, lip biting, and nervous laughter" (Baumrind 1985; Shanab and Yahya 1977). Indeed, replications and variations of Milgram's study continue to be done (Perry 2012; Burger 2009; Burger, Girgis, and Manning 2011), and what is puzzling about these contemporary attempts at replication is their avoidance of the question Baumrind focuses on: whether or not present-day participants are obedient to the same degree as participants in Milgram's original study, are these participants continuing to be harmed by participating? It is, after all, not just an interesting scientific question whether participating in a Milgram-style obedience study is traumatic, even though the existence of such trauma is an important piece of data.

Baumrind argues that because the APA's code did not block replications of Milgram's study, it was unable to protect participants in the relevant studies from harm. It is important to note, furthermore, that Baumrind's argument is not only that the code must be revised because it is not adequately protecting participants in such studies. Rather, the argument is also that any philosophical or legal use of the code can fail to adequately capture the costs of research to participants, such as when the people who are applying the code are morally ignorant in certain predictable ways, and where this ignorance is a by-product of professional training. Said differently, a psychologist trained only in laboratory research cannot calculate the gains in scientific knowledge that balance out or compensate the costs to participants of every type of potentially harmful research—because the harms participants experience cannot be described using the conceptual vocabulary that can be used to describe gains to scientific research. Knowing how to speak about experimental research in the language of social psychology is quite different from knowing how to use a moral language to describe the psychic harms caused to people by the very same research. The two different vocabularies will be incommensurable, and so it is an elementary fallacy to claim, baldly, that they balance one another out. As Baumrind writes, "It takes well-trained clinical interviewers to uncover true feelings of anger, shame, or altered self-image in participants who believe that what they say should conform with their image of a 'good subject'" (Baumrind 1985, 168). Thus, Baumrind concludes that the configuration formed out of the methodological norms of psychological experimentation and the prescriptive norms of the APA's *Ethical Principles* functioned to cause people harm. Changing both the methodological norms and the more formal ethical norms would therefore be a way of causing better outcomes.

Public Policy and Education in Bangladesh and Tennessee

The largest NGO in the world is BRAC. Now a global organization, BRAC makes its home in Bangladesh, where its rural primary and preprimary education programs have been some of its most successful activities. Of particular note is the fact that the method BRAC employs for developing its school policies is quasi-experimental. The NGO opens a number of schools and explicitly formulates the basic norms used to structure the educational day, which are then selectively modified with the aim of producing a particular cluster of good educational outcomes. BRAC researchers Mirja Mohammad Shahjamal and Samir Ranjan Nath offer an example of this process:

Before initiating pre-primary education programme with a full swing, BRAC experimented a number of issues on a trial and error basis from 1997 to 2001. The experiment included following questions to be answered.

1. Who should be the prospective teachers for pre-primary classes?
2. Which section of the children should be served by the programme in terms of age and social stratification?
3. What is the future prospect of such [a] programme?
4. What are the strategies need[ed] to be taken for [the] pre-primary programme?

At the beginning of this experiment, BRAC appointed female graduates of BRAC non-formal primary schools who at the same time were the students of the formal secondary schools. ... The teachers worked as a volunteer and no remuneration was provided to them. After one year, it was understood that one teacher is inadequate to manage the students. The concept of two teachers came in to concentrate more on students' learning. Although the teachers were better able to handle the pupils' problems, it was found that conceptual and developmental aspects of the students were not reaching to the satisfactory level that BRAC wanted. This lead BRAC to consider a bit higher educated women as teachers instead of BRAC school graduates. (Shahjamal and Nath 2008, 9)

BRAC has received a fair amount of praise for the success of its programs. By the lights of the causal theory of ethics, moreover, BRAC has identified some of the norms that should be regarded as ethical in the normative context of rural education in Bangladesh. Crucially, these norms have been developed through the implementation of feedback loops running between policy (the structural content of regulative frameworks) and practice (measures of the effects of realizing different regulative frameworks):

Monitoring is considered a crucial element for improving the programme's quality. One supervisor usually monitors 20–25 primary schools and 10–15 pre-primary schools selected at random. Evaluations focus on both the qualitative and quantitative aspects of the programme. In order to ensure the quality of the evaluation results, standardized guidelines are provided for monitors. Monitors attend all classes and subjects on a given day in order to assess classroom-based teacher-student interactions and the delivery of lessons. The results are shared with the teachers who will, in turn, discuss any problems identified with the students and take corrective measures. (UNESCO Institute for Lifelong Learning 2013)

The processes described here ultimately lead to revisions of norms, big and small, that structure the classroom such that the theoretical content of the norms develops more or less in equilibrium with the ability to increasingly realize the patterns of education that are the desired outcome of BRAC's education program. Yet these norms are inherently local. They "work" to produce good outcomes in Bangladesh, and perhaps nowhere else—although the BRAC program has been successfully implemented in a number of other countries, including Uganda and Pakistan. The point, then, is that BRAC has discovered which norms cause a particular valuable outcome in rural Bangladesh; BRAC has not discovered what general norms are universal causes of effective education, and, almost certainly, there are no such universal norms.

The Tennessee Class Size Project—sometimes called the STAR Project (Word et al. 1990)—is a famous example of a similar study on the formation and effects of educational policies. The project specifically studied the effects of class size on both short-term and long-term student performance, and the study spanned seventy-nine schools in rural, semi-rural, and metropolitan areas. The project ran from 1985 to 1989 and collected data on the outcomes for students in smaller class sizes, in regularly sized classes with teacher's aides, and in regularly sized classes without teacher's aides. The classes were at the kindergarten through grade 3 levels. The performance of students was also tracked once they left the study and returned to regularly sized classrooms in grades 3, 4, and 5. The data produced by the study are detailed, and as is usually the case for such data, they warrant no simplistic interpretations. However, generally speaking, the study did show that smaller-sized classes did improve the performance of children on subsequent tests, and that this benefit persisted once the students returned to regularly sized classes (Mosteller 1995, 125). The point here is the same as with BRAC: arguably, this research has discovered a cluster of ethical norms for normative contexts inherent to primary education in Tennessee—meaning that these norms should be fundamental to the extent possible in the context of educational policy. These are, of course, the norms that create "smaller" classroom sizes, as they do better in producing a good outcome than norms that create regular classroom sizes.

Ethical Norms and Laws

Not all laws are directly connected to causing good or valuable social outcomes; but sometimes it is possible to discover which laws have this kind of connection. And when such discoveries are made, this too counts as answering a hard question. A provocative example of this is provided by Scott Cunningham and Manisha Shah's research on prostitution in Rhode Island (Cunningham and Shah 2014). Summarizing their findings, they write,

> We exploit the fact that a Rhode Island District Court judge unexpectedly decriminalized indoor prostitution in 2003 to provide the first causal estimates of the impact of decriminalization on the composition of the sex market, rape offenses, and sexually transmitted infection outcomes. Not surprisingly, we find that decriminalization increased the size of the indoor market. However, we also find that decriminalization caused both forcible rape offenses and gonorrhea incidence to decline for the overall population. Our synthetic control model finds 824 fewer reported rape offenses (31 percent decrease) and 1,035 fewer cases of female gonorrhea (39 percent decrease) from 2004 to 2009.

This is not evidence that prostitution should be legalized to the extent that it is allowed to occur indoors. Rather, it is an example of a natural experiment that provides some insight into the effects of legal norms on ethical outcomes for people. Further research here, then, may lead to insight into how to define laws regulating or outlawing prostitution so that rape and the transmission of sexual disease are ameliorated—both good outcomes.

Hard-Question Research and Metaphysical and Epistemic Progress in Ethics

We now have five different examples of different kinds of social theory that can generate answers to hard questions, and in chapter 9, we will explore how this kind of social research can be used to influence the direction of research in moral psychology. The remaining work for this chapter is to answer two objections to the line of argument developed so far.

The first objection: if hard-question social research is so common, then why is it not easier to observe clear-cut cases of ethical progress? The answer to this question rests in a distinction that helps prevent overinterpreting the ethical importance of hard-question social research. This is the distinction between *epistemic progress* and *metaphysical progress* in ethics. Epistemic progress can be realized by increases in the accuracy, breadth, content, usefulness, and coherency—in short, increases in the epistemic virtues—of ethical knowledge. Metaphysical progress can be realized by changes in the social world that increase fairness, or compassion, or liberty—in short, increase in both the number and the mutual causal integration of good and valuable social outcomes. So far as only the causal theory of ethics is concerned, epistemic progress will come in the form of insights into which particular clusters of norms either do, or would, cause good social outcomes, although, technically, epistemic progress can also be achieved by developing insights into which regulative frameworks cause human misery and suffering. Knowing the causes of bad outcomes is itself a form of ethical knowledge, which is why it is possible to make epistemic progress by learning more about how to increasingly effectively cause, say, widespread immiseration or poverty. But metaphysical progress in ethics can only be made along a single dimension, because metaphysical progress just is better stabilizing and balancing existing valuable outcomes for a normatively integrated population, while also discovering how to add new good outcomes to the existing network of good and valuable outcomes. The two kinds of progress are related insofar as epistemic progress in ethics is necessary but not sufficient for metaphysical progress: insight into the causes of good or bad outcomes will be useful, in part, for figuring out how to ameliorate the bad outcomes. Nevertheless, this knowledge will rarely if ever be sufficient to design the democratic proposals—or policy changes, or new incentive structures, or inspired literary works, or patterns of education, or systematic philosophical arguments, and so on—that can each in their own way help to bring about metaphysical change.

Socially Realistic Normative Theory and Its Relevance to Moral Psychology

The next objection is more substantial, and consequently a certain amount of foregrounding is necessary to fully appreciate its force. The five examples of hard-question social theory that we have examined show the different ways in which investigations into the structural content of social institutions can lead to particular insights into the causes of good and valuable outcomes. The causal theory of ethics stipulates that these insights should be turned into (defeasible) discoveries about which clusters of norms are ethical norms. But it also predicts that these insights will almost certainly not coalesce into a universal theory of the causes of valuable outcomes that transcends differences between normatively integrated populations. The causal theory of ethics is designed to accommodate the fact that knowledge of how norms cause good outcomes is inherently local knowledge, and as it concerns the methodology of moral psychology, this is a very important consequence of this feature of the causal theory. There will not be a single set of concepts or definitions that are sufficiently generally applicable to be useful for isolating the moral domain as it exists in all different normatively integrated populations around the world; and also have a degree of specificity allowing them to pick out within these populations the local norms that, per the causal theory of ethics, are the distinctly ethical norms. Moral psychology therefore should begin from social theory that has a particular solution or is developing a solution to a specific hard question, and aims initially for tests of local external validity only (as I will argue in chapter 9). For now, the point is simply that moral psychologists should not try to study ethical behavior as a sui generis construct; instead, they should focus on local and quite possibly unique patterns of moral behavior. Moral psychologists can do this by using the causal theory of ethics and hard-question social theory to locate the relevant patterns of behavior.

That is, however, a difficult methodological implication to grasp when it is expressed as an abstract claim about the referential properties of concepts useful for structuring scientific practice. But there is an easier route to the same conclusion that allows us to contrast the causal theory of ethics with an alternative collection of normative theories in economics. For the objection to be considered in this subsection is this: What justifies adopting the causal theory of ethics as the guiding ethical theory for research in moral psychology, when economics provides us with several better-known and much more theoretically developed alternatives? The answer, in short, is that because the causal theory of ethics focuses on local norms and their effects, it is more socially grounded than the alternative normative theories in economics.

To further explain: in the sciences, knowledge of causal structures tends to occur in one of three forms. The first of these forms is the formulation, introduction, revision, and elimination of different concepts, all with the specific aim of producing a conceptual vocabulary that is able to generate multiple different inductively or explanatorily useful descriptions of the relevant causal properties. Call this the *causal descriptive stage*. Perhaps the most famous example of this kind of scientific achievement is Darwin's *On the Origin of Species* (Darwin [1859] 2008). Darwin introduced no universal laws that purported to capture perfectly general biological regularities, but he did define a set of technical concepts that could be used to formulate specific patterns of explanation, and he showed these explanatory patterns could be used to make sense of observations from a very broad number of different scientific fields.

The second form of knowledge that scientists produce comes in the form of models of the relevant processes. Sometimes properties of the relevant causal processes that can be described by the earlier concepts are nevertheless not referred to by any of the terms of the model, and where this simplification and intentional omission is done, it is done because it allows us to construct models that provide us with deeper insight into the counterfactual-supporting interactions of the relevant phenomena. The point is that models often trade descriptive breadth for predictive precision and causal explanatory power (Giere 2009; Norton 2012; Bokulich 2011). Furthermore, as we have noted several times, it is not uncommon for a scientific field to reach a point at which it is using mutually inconsistent models to describe one and the same complex causal system, where the usefulness of a model is partly determined by the explanatory ends of the scientists trying to learn about the causal system in question. These models permit scientists to represent a system's most important interactions and behaviors, how it changes as the result of certain interventions and not others, and, finally, what the most important causal components of the system probably are. These various different forms of empirical knowledge often come at a cost of accepting, or at least choosing to work with, a model with known simplifications, fictions, and errors—which is also why the respective inconsistencies between the models is not a serious issue. Call this kind of scientific achievement the *approximate-but-useful model stage*.

Very often a collection of useful and approximately true models—the approximate-but-useful model stage, again—is the practical limit of scientific inquiry, and this limit is enforced simply by the interaction between the complexity of the causal systems we are trying to understand and the limited epistemic capacities of the human mind. Even in extremely applied and highly automated contexts, like civil and aeronautical engineering, scientists still must rely upon collections of approximate models, like those used in computational fluid dynamics.

However, in some circumstances it is possible to advance one stage further, to what can be called the *precise formal model stage*. This kind of achievement is characterized by the production of formal theories with explicitly defined rules of inference, where the inferences are almost exact representations of specific causal transformations that the system can go through, and where the terms of the relevant formalism are known to be almost exact representations of the underlying causal properties (magnitudes, forces, etc.). Usually this formal precision comes at a cost of generality: the phenomena that can be described by scientific theories produced at the precise formal model stage are normally human-made phenomena (Cartwright 1983). Examples of this kind of research would be the use of balanced equations in analytical chemistry, or the use of electronic design automation in microprocessor design; these formal models apply to specimens, or experimentally isolated phenomena, or data sets that have been "cleaned" or are otherwise abstracted away from some broader and more complex set of information. Scientific precision is often man-made.

Most scientific theories are probably at either the causal descriptive stage or the approximate-but-useful model stage, and it should be clear that the causal theory of ethics is formulated with this general tendency in mind. According to the causal theory of ethics, the limits of reliable ethical inquiry will often be only causal descriptions that provide information about the links between local institutions and specific patterns of valuable outcomes. In some very specific contexts it may be possible to construct mathematical models that describe some of the more complex causal interactions, but there is no presumption built into the causal theory of ethics that by applying it to learn about which norms are and are not ethical, theorists will somehow come to learn the universal laws of social reality. (And as an interesting side note, this is one of the reasons why, in pursuing a socially realistic ethical theory, it is important to not conceptualize ethical norms as categorical truths or as context-insensitive necessary principles, for doing so is a way of cutting the concept of an ethical norm free from the complexity of reality of human social behavior, and thereby rendering ethical theory implausible from the perspective of contemporary scientific practice.)

The problem we are moving toward addressing is whether a moral psychologist should choose the causal theory of ethics over some form of economic utilitarianism, since, again, some may object that economic utilitarianism should, for our specific purpose, be preferred to the causal theory of ethics on the grounds that, for example, it has a vastly deeper history of scientific development and successful application. The response, however, is that that the models of economic utilitarianism are too abstract to be as useful as the causal theory of ethics to moral psychology—and the technical reason for this is that most of these models are presented as though they are at the precise formal model stage of scientific development, even though there is sometimes only very little evidence that they have achieved even the causal descriptive stage. Dani Rodrik puts the same point more simply when he notes that economists have a tendency to "confuse a model for *the model*" (Rodrik 2015, 144). This is a problem for economic utilitarianism's usefulness for moral psychology because it prevents the relevant theories from specifying in a sufficiently fine-grained way which social behaviors are in fact the ethical behaviors, and which behaviors are not.

Let me now explain why that is the case. The theories in economics that are most relevant alternatives to the causal theory of ethics—because they are both philosophically and scientifically plausible, and at least ostensibly, about real patterns of social behavior—are the versions of utilitarianism developed by economists in the twentieth century (e.g., Becker, Murphy, and Werning 2005; Stiglitz 1982; Mirrlees 1982; see also Posner 1979 for commentary), most notably in welfare economics (Samuelson 1983; Hicks 1939; Harsanyi 1953), its contemporary descendant. Tellingly, George

Stigler notes the birth of this field of research, writing, "Quite recently there has been a return to the view that the treatment of welfare problems is an integral part of economic analysis. The new welfare economists claim that many policies can be shown (to other economists?) to be good or bad without entering a dangerous quagmire of value judgments" (Stigler 1943, 355). The theories that followed afterward together usually have two technical assumptions that allow for a great deal of mathematical creativity—namely, normative problems are treated as problems that have unique, discrete solutions (such as when we can maximize some value or satisfice some arrangement of preferences, and so on), and that the domain of application is universal, and so the relevant calculation problems can be formulated either generically or abstractly (Backhouse 2014). And then there is a third methodological assumption that is common to both economic utilitarianism and a significant amount of contemporary normative philosophy. Since the normative theories are prescriptive in nature, this means that there is no need to introduce into the methods used to determine the truth of the relevant model or formalism a methodological feedback loop running both back and forth from data derived from existing empirical circumstances to the details of the model. Instead, it is the obligation of the world (humans, that is) to conform to the model, and not the model to find a way of fitting the world. Consequently, model fitting becomes a vastly less important exercise in both economics and the corresponding areas of normative philosophy than in other social and natural sciences that make use of models, and when a model fails, it is not uncommon to interpret its failure as a by-product of bad policy, democratic mistakes, and not inaccurate modeling ideas.

Now, there are some structural phenomena in some larger societies that are very well approximated by some of the relevant models. Aggregate demand is likely well captured by, for instance, representative agent models used by central banks (Hartley and Hartley 2002). But that really is not the issue here. The only question that is important is whether or not these models can do as well as the causal theory of ethics in specifying which patterns of existing behavior are ethical, or at least "normatively appropriate" given some ethically valuable ends, in a way that is more useful for moral psychology than the causal theory of ethics. And there is a metaphysical test that we can apply to resolve our question: is the complexity of social reality such that one cannot theorize about it in perfectly general, perfect abstract terms, and so must instead make do with a collection of motley and constantly growing models and causal theories that cannot themselves be derived from a small set of fundamental axioms—and which are true only of specific historical, cultural, or geographic areas and periods? Or is social reality a domain with enough law-like causal regularities that it is possible to use formal theories with a small number of general axioms to describe its fundamental processes and patterns of transformation? If it is the latter, then economic utilitarianism can save the social sciences a lot of fieldwork in delimiting the ethical dimensions of these worlds; if it is the former, then economic utilitarianism is too abstract to capture any existing patterns of behavior in the detail needed for the models to be useful to moral psychology.

To determine which answer is true, we ask: Is it *morally appropriate* to assume that the social world is simple, and has law-like regularities? It seems that it is not. For a vivid illustration of this point, here is M. Ali Khan's assessment of Peter Singer's recent defense of his (Singer's) version of economic utilitarianism:

> Peter Singer's *One World: The Ethics of Globalization* is, to me, an unwelcoming work, and at one level, this essay is simply an effort to understand my own puzzlement, if not fear and anxiety, at his vision of a single world. His appeal to a universal solvent of reason and of rationality, and his application of it to dissolve what he refers to as *moral relativism*, strips me of what I see to be my sense of self and my basic dignity. It is a picture of a world that is to be relentlessly homogenized and beaten into shape by the market, by utilitarianism, by the internationally powerful, and it is to be done in a style that demands my gratitude for all that is being done, remains to be done, to my beliefs, to my definitions, to me. (Khan 2004, 363)

To fully grasp the meaning of Khan's remarks, it is important to understand that he is among the most preeminent mathematical economists alive. He understands as well as anyone what Singer advocates when he describes one world under one law, and where this law is roughly an extension of the WTO's trade and investment policies. Yet Khan observes that Singer is not sensitive to even the most basic properties that individuate different realized conceptions of the good and valuable—he notes that Singer, for example, "does not mention the Quran, does not show any awareness that Muhammad, may he rest in peace, at least according to a non-negligible set of people on (what the text refers to as) 'this planet,' is not the author of the Quran, and that his teachings, as embodied in the Sunnah, also have authorial references" (Khan 2004, 273).

Now, we usually individuate norms by reference to their propositional content. But Khan's insight shows that two (clusters of) norms with exactly the same propositional content can give rise to very different motivational dispositions, depending on what other norms are accepted as part of existing realized regulative frameworks, and so estimating or predicting the behavioral effects of adopting a small number of norms requires more information than just the propositional content of the norms themselves. For example, the norm *choose minimally Pareto efficient arrangements* and the norm *maximize welfare while minimizing suffering* may seem to be culturally neutral, but they are not. Their effects on behavior, if and when they become embedded in the structural content of existing regulative frameworks, will depend a great deal on what other norms are also part of the structural content of the relevant frameworks. If these new norms are imported into a cultural context in which they conflict with existing normative arrangements, then this really is to strip certain people of their normative identity; that is the moral objection. But going in the other direction, without a sense of the existing regulative frameworks of a particular normatively integrated population, a theorist who knows only the abstract formulations of certain norms will be unable to reliably estimate the causal effects of their realization, or whether there is a pathway of possible cultural development from the existing regulative frameworks (that is, social institutions) to the proposed new set of frameworks. So, the moral argument shows both that it is important not to aim for normatively neutral models of the causes of good and valuable outcomes, as when, for instance, economists treat a social world as composed entirely of sets of optimization problems (Dorfman, Samuelson, and Solow 2012, chap. 14; Franklin 2002), and that, more technically speaking, models formulated at that level of abstraction will not contain enough causal information about existing patterns of thought and behavior to provide any useful guidance to a moral psychologist who faces the problem of locating the ethical behaviors in a population. Herbert Gintis puts very nearly the same idea this way when he writes that "Neoclassical [welfare] theory . . . takes preferences as either fixed, or changing only in response to variables external to the model. In positive economics, the formation of preferences is relegated to sociology or social psychology; and in welfare economics, preference structures are among the fundamental, unexplained data" (Gintis 1974, 415).

The causal theory of ethics should therefore be preferred because it is able to provide a moral psychologist with more accurate guidance about where to find patterns of ethical behavior in different social worlds. That is because, by design, it hews more closely to the causal descriptive stage of scientific knowledge—about preferences expressed as conceptions of the good and valuable, and about the causal links between regulative frameworks and valuable social outcomes—than theories in economic utilitarianism.

Conclusion

We now have the components of a solution to the problem of inconsilience between ADJDM moral psychology and philosophical ethics. The causal theory of ethics is a philosophical theory that provides a conceptual framework that provides guidance about which (and certainly not *only* which) social behaviors are ethical behaviors. "Hard-question" research in the social sciences shows when, where, and among whom the relevant behaviors are realized—and so, technically, this kind of social theory yields knowledge of normative contexts.

Crucially, both the causal theory of ethics and hard-question social theory will typically be conceptually and methodologically autonomous from any particular area of psychological inquiry. That is not to say that either the probability of hypotheses about which norms are ethical (as per the causal theory of ethics) or hypotheses about which social behaviors cause good outcomes (as per hard-question social science) will be *independent* of the probability of any number of psychological hypotheses. It is only the claim—mirroring our discussion of the relationship between ultimate explanations and proximate explanations in chapter 2—that it will not be possible to derive, for instance, answers to "hard questions" from theories about the psychological representations of structural content (because the latter theories will elide population-level details about the realization of particular valuable outcomes). Likewise, it will not be possible to derive theories about which beliefs constitute the psychological representation of a regulative framework from theories about how the social institution formed by the regulative framework contributes to a host of positive social outcomes (because the latter theories will normally exclude reference to any specific psychological state). More simply put, the relationship between plausible theories about the effects of regulative frameworks on good and valuable outcomes and plausible theories of the psychological representations of the structural content of the relevant regulative framework will be many-to-many.

That all means the framework provided by the causal theory of ethics paired with specific answers to hard questions can generate methodological and metaphysical consilience among ethical inquiry, inquiry in evolutionary biology, and empirical research in moral psychology—if, of course, the information provided by the framework formed out of the causal theory of ethics plus hard-question social theory can be turned into research questions that can be answered by moral psychology. Showing how this can be done is the task of the next chapter.

References

Backhouse, Roger. 2014. "Revisiting Samuelson's Foundations of Economic Analysis." *Social Science Research Network*, November. doi:10.2139/ssrn.2510383.

Baumrind, D. 1964. "Some Thoughts on Ethics of Research: After Reading Milgram's 'Behavioral Study of Obedience.'" *The American Psychologist.* http://psycnet.apa.org/journals/amp/19/6/421/ (accessed July 10, 2014).

Baumrind, D. 1985. "Research Using Intentional Deception. Ethical Issues Revisited." *The American Psychologist* 40 (2): 165–174.

Becker, Gary S., Kevin M. Murphy, and Iván Werning. 2005. "The Equilibrium Distribution of Income and the Market for Status." *The Journal of Political Economy* 113 (2): 282–310.

Bokulich, Alisa. 2011. "How Scientific Models Can Explain." *Synthese* 180 (1): 33–45.

Burger, Jerry M. 2009. "Replicating Milgram: Would People Still Obey Today?" *American Psychologist* 64 (1): 1–11.

Burger, Jerry M., Zackary M. Girgis, and Caroline C. Manning. 2011. "In Their Own Words: Explaining Obedience to Authority Through an Examination of Participants' Comments." *Social Psychological & Personality Science* 2 (5): 460–466.

Cartwright, Nancy. 1983. *How the Laws of Physics Lie.* 1st ed. Oxford: Clarendon Press.

Cunningham, Scott, and Manisha Shah. 2014. "Decriminalizing Indoor Prostitution: Implications for Sexual Violence and Public Health." NBER Working Paper

20281. National Bureau of Economic Research. http://papers.nber.org/tmp/55308-w20281.pdf (accessed March 10, 2015).

Darwin, C. [1859] 2008. *The Origin of Species*. New York: Random House Publishing Group.

Dorfman, R., P. A. Samuelson, and R. M. Solow. 2012. *Linear Programming and Economic Analysis*. Mineola, NY: Dover Publications.

Franklin, J. N. 2002. *Methods of Mathematical Economics: Linear and Nonlinear Programming, Fixed-Point Theorems*. Philadelphia: Society for Industrial and Applied Mathematics.

Giere, Ronald N. 2009. "An Agent-Based Conception of Models and Scientific Representation." *Synthese* 172 (2): 269–281.

Gintis, Herbert. 1974. "Welfare Criteria with Endogenous Preferences: The Economics of Education." *International Economic Review* 15 (2): 415–430.

Harsanyi, John C. 1953. "Cardinal Utility in Welfare Economics and in the Theory of Risk-Taking." *The Journal of Political Economy* 61 (5): 434–435.

Hartley, J. E., and J. E. Hartley. 2002. *The Representative Agent in Macroeconomics. Routledge Frontiers of Political Economy*. New York: Taylor & Francis.

Hicks, J. R. 1939. "The Foundations of Welfare Economics." *The Economic Journal* 49 (196): 696–712.

Khan, M. A. 2004. "Regional (East-West/North-South) Cooperation and Peter Singer's Ethics of Globalisation." *Pakistan Development Review* 43 (4): 353–396.

Lane, R. E. 2000. *The Loss of Happiness in Market Democracies. The Institution for Social and Policy St Series*. New Haven, CT: Yale University Press.

McWhorter, L. 2009. *Racism and Sexual Oppression in Anglo-America: A Genealogy*. Bloomington: Indiana University Press.

Milgram, S. 1963. "Behavioral Study of Obedience." *Journal of Abnormal Psychology* 67 (October): 371–378.

Mirrlees, James A. 1982. "The Economic Uses of Utilitarianism." In *Utilitarianism and Beyond*, ed. Amartya Kumar Sen and Arthur Owen Williams Bernard, 77–81. Cambridge UK: Cambridge University Press.

Mosteller, F. 1995. "The Tennessee Study of Class Size in the Early School Grades." *The Future of Children*. Critical Issues for Children and Youths 5 (2): 113–127.

Norton, John D. 2012. "Approximation and Idealization: Why the Difference Matters*." *Philosophy of Science* 79 (2): 207–232.

Perry, Gina. 2012. *Behind the Shock Machine: The Untold Story of the Notorious Milgram Psychology Experiments*. Melbourne: Scribe Australia.

Posner, Richard A. 1979. "Utilitarianism, Economics, and Legal Theory." *The Journal of Legal Studies* 8 (1): 103–140.

Rabinow, Paul, and Nikolas Rose. 2006. "Biopower Today." *BioSocieties* 1 (2): 195–217.

Rodrik, Dani. 2015. *Economics Rules: The Rights and Wrongs of the Dismal Science*. 1st ed. New York: W. W. Norton & Company.

Samuelson, Paul Anthony. 1983. *Foundations of Economic Analysis*. Cambridge, MA: Harvard University Press.

Shahjamal, Mirja Mohammad, and Samir Ranjan Nath. 2008. *An Evaluation of BRAC Pre-Primary Education Programme*. Dhaka: BRAC.

Shanab, M. E., and K. A. Yahya. 1977. "A Behavioral Study of Obedience in Children." *Journal of Personality and Social Psychology* 35 (7): 530–536.

Stigler, George J. 1943. "The New Welfare Economics." *The American Economic Review* 33 (2): 355–359.

Stiglitz, Joseph E. 1982. "Utilitarianism and Horizontal Equity." *Journal of Public Economics* 18 (1): 1–33.

UNESCO Institute for Lifelong Learning. 2013. "Effective Literacy Practice." *LitBase*, May. http://www.unesco.org/uil/litbase/ (accessed May 1, 2015).

Word, Elizabeth, John Johnston, Helen Pate Bain, B. DeWayne Fulton, Jayne Boyd Zaharias, Charles M. Achilles, Martha Nannette Lintz, John Folger, and Carolyn Breda. 1990. *The State of Tennessee's Student/Teacher Achievement Ratio (STAR) Project*. Nashville: Tennessee State Department of Education.

9 Descriptive Moral Psychology

This chapter proposes that any line of psychological research that focuses its attention primarily upon the psychological representations or physical presuppositions of regulative frameworks that produce patterns of ethical behavior should be called *descriptive moral psychology*. Then, in the chapter 10 we will examine the details of a complementary form of research, *normative moral psychology*, which focuses on enabling knowledge.

But there is an important additional area of concern for descriptive moral psychology. Recall the GapMinder data presented at the beginning of chapter 6. The charts we examined represent changes in the average amount of two types of valuable outcomes, health and wealth, across many different populations. It is an elementary observation to note that most people's access or enjoyment of these two outcomes will be different than the average value for their population. Of course, research at the structural level of analysis can estimate the shape of the distribution around a central tendency, and from these estimations we can make some a priori inferences about likely patterns of nontypical outcomes. Yet it is immensely important to go much further than this, to try to understand the diversity of valuable outcomes that are otherwise hard to describe in conceptual vocabularies apt for structural explanations and analyses fitted to the level of large-scale normatively integrated populations. The point is that there are, intuitively, good outcomes at this level of social behavior that are "too small," "too scattered," and "too specific to a very localized groups of people" to show up in population-level analysis of human behavior. Nor are these goods that are direct, functional by-products of social institutions; they are not the good outcomes that any social institution is *designed* to produce. And yet, these goods are more than just personal outcomes that mesh only with a single person's conception of the good *just for me*. A secondary group of questions for descriptive moral psychology, then, aims to investigate the diversity both of different conceptions of the good and valuable and of corresponding different successes or failures in outcomes, where both are patterns of conception and outcome that are not well described by even the most rigorous hard-question social research. And in many cases these will be patterns of thought, behavior, and outcomes that occur at what Bronfenbrenner (1979) termed the "microsystem," which is the world of families, friends, school peers, and others as small-scale, person-to-person interactions.

In the following sections I describe some of the basic questions and methodological ideas motivating two different research programs that can be seen as examples of descriptive moral psychology. These remarks are intended to illustrate two rough philosophical models of what it might look like for researchers to begin working out an empirical understanding of the psychological foundations of some real-life ethical behaviors.

What's New Here?

A natural place to start our investigation into descriptive moral psychology is with the question: What's new here? What makes descriptive moral psychology a new research direction in moral psychology? The answer, in short, is that descriptive moral psychology focuses on the patterns of cognition closely associated with real patterns of social behavior, and because of that, it does not intentionally aim to develop psychological theories that will likely have universal external validity. By contrast, ADJDM moral psychology most often focuses on comparatively "deep" or "foundational" psychological mechanisms that, if real, explain a very large number of social behaviors, and because of that, ADJDM moral psychology characteristically produces theories that aim for validation by tests of universal external validity—despite the fact that typically no efforts are made to assess even the local external validity of the relevant hypotheses.

An example can help make this clear. In a recent experiment, Philip Pärnamets and colleagues report successfully manipulating people's decisions about the truth or falsity of abstract moral propositions in a two-option, forced-choice probe that used length of gaze as the independent variable. Here is how they summarize their study:

We monitored participants' eye movements during a two-alternative forced-choice task with moral questions. One option was randomly predetermined as a target. At the moment participants had fixated the target option for a set amount of time we terminated their deliberation and prompted them to choose between the two alternatives. Although participants were unaware of this gaze-contingent manipulation, their choices were systematically biased toward the target option. We conclude that even abstract moral cognition is partly constituted by interactions with the immediate environment and is likely supported by gaze-dependent decision processes. By tracking the interplay between individuals, their sensorimotor systems, and the environment, we can influence the outcome of a decision without directly manipulating the content of the information available to them. (Pärnamets et al. 2015)

Obviously, the interplay of individuals, their sensorimotor systems, and the environment influences choices, moral or otherwise; that interaction almost never is *not* one of the factors influencing decision making. For our purposes, thus, the only relevant result is that a low-level psychological phenomenon, gaze fixation, was observed to be associated with patterns in hypothetical moral choices.

And, to be clear, the reason I am bringing this study into the discussion is not to criticize the ecological validity of the study. Rather, I want to pursue a slightly different line of analysis. Since there is no such thing as a perfectly generic, perfectly decontextualized moral choice, do the authors of the study have a view about which—some, all, few, or many—real-world moral choices are influenced by fixation patterns? It would seem that "all" is the answer most compatible with Pärnamets and colleagues' interpretation of their own data:

In simple words, we find that even for moral choices we end up preferring what we fixate. Future work should investigate and model the computational properties underlying moral decisions directly to facilitate the necessary model comparisons. Regardless, our results demonstrate that moral choices, by virtue of their embodied dynamics, fall within the scope of traditional accounts of the computational processes underlying human decisions. This suggests future prospects for developing domain-general decision models covering choices from deciding the direction of randomly moving dots to the legitimacy of euthanasia. (Pärnamets et al. 2015)

Obviously, the legitimacy of euthanasia is not itself a generic problem. The construct *euthanasia* is defined very differently across disparate medical, social, and legal frameworks; and the philosophical point here is that these differences are simply outside of the scope of analysis of Pärnamets et al. There really is no evidence in the study that gaze fixation patterns are possible explanations of even just one or two real, and thus contextually specific, ethical decisions.

This example mirrors one of the deeper intellectual tendencies in ADJDM moral psychology, namely, that very few researchers make any meaningful attempt whatsoever to *test* whether or not some independently observed social behaviors are explained by the psychological mechanisms described in the research literature. That is true despite confident predictions about the range of choices and behaviors apparently explained by relevant psychological theory—for a striking example of the trend, see Phillips and Shaw 2015. Tage Rai and Alan Fiske put the same underlying observation this way: "by adopting this distinction between moral psychology per se and the social influences that distort moral judgment (...) moral psychology largely separated [itself] from social psychological studies of prescriptively immoral real-world behaviors and anthropological findings regarding diverse moral practices across cultures" (Rai and Fiske 2011, 58).

So, what would it look like to test whether or not a hypothesis in moral psychology explains some real world moral decisions? As it pertains to the Pärnamets et al. study, such a test would be, for instance, a follow-up study in which participants who have developed beliefs about euthanasia are drawn from hospital staff in Botswana, Tibet, Canada, and Italy, and where the research question is whether or not participants in such a study also display the fixation-choice response. There are two scientific results that would likely be produced by such a study. First, even if the fixation-choice association appeared only in, say, the Canadians, then the country-by-country variation in the fixation-choice association would be an interesting observation. Second, and more importantly, it would be a striking result if records of real-life decisions about euthanasia made by the study participants were (or were not) correlated with observed associations between fixation and choices about euthanasia. And the more general point that can be made here is that the Pärnamets et al. study is important because it does identify one psychological mechanism—indeed, a cognitive mechanism that is present in nearly every human—that could be part of the psychological explanation of a very large number of different social behaviors. Maybe patterns in gaze duration and fixation really do explain different behaviors that, in turn, yield different good outcomes in different social worlds. Indeed, that is exactly the question to ask for descriptive moral psychology: do psychological mechanisms $PM_1 \ldots PM_n$ explain any of the behaviors that we know are the causes of good outcomes?

Thus, what makes descriptive moral psychology new—what makes it an advance from conventional ADJDM research in moral psychology, like the Pärnamets et al. study—is that by design it places a premium on tests of whether psychological hypotheses are able to explain real-world social behavior. And so, the crucial methodological change would be to withhold assent to conclusions—such as (to quote Pärnamets et al. one last time) "the process of arriving at a moral decision is not only reflected in a participant's eye gaze but can also be determined by it"—until some real-world ethical behaviors (or decisions leading to the relevant behaviors) are shown to be causally linked with, in this case, patterns in gaze fixation.

The Container Test

Descriptive moral psychology can be carried forward by pursuing, among many research programs, either of two programs detailed here. The first of these, the container test, gets its name from the more than slightly metaphorical idea that we can test the local external validity a theory in descriptive moral psychology by examining whether or not the causal processes described by hard-question social theory can "contain" the relevant psychological model. The idea, in other words, is to examine whether previously observed ethical behaviors can be explained by a psychological theory, and if they can, then how many aspects of the relevant behavior are accounted for by the relevant psychological theory.

One of the benefits of linking research in moral psychology with hard-question social theory is that this link will provide guidance about which populations researchers doing descriptive moral psychology should sample their participants from. Randomly selected undergraduates at research universities are not always a good population to sample from for psychological studies intent on reaching conclusions about moral psychology—and not just because attending university changes a person's normative outlook (Gurin et al. 2002). Rather, the student body of a research university is commonly an inherently cosmopolitan group, and the students are a transient population who are members of the university community for only brief periods of time. This point adds to the now famous argument developed by Richard Heinrich and his collaborators that most participants in psychological studies are Western, educated, industrialized, rich, and democratic—in short, WEIRD (Henrich, Heine, and Norenzayan 2010). Or, to put the same point in a more general way: the more historically and culturally local the realizations of ethical thought and behavior are, the more important the question becomes of which population to sample from for the purposes of doing moral psychology. Assuming—with Kant—that any human whomsoever has roughly the same moral knowledge as any other, and assuming in conjunction with this—and with Mill—that all moral dilemmas have rational solutions yields the implication that almost any rational person will be as good as any other at solving the problems, and therefore that any such person is as appropriate as any other for serving as a participant in a moral psychology study. But these two assumptions may be false; certainly they should not be a priori methodological axioms in moral psychology, just it should not be an a priori axiom in the methodology of experimental psychology that WEIRD people are appropriate participants for any psychological study aiming to generalize its conclusions as broadly as possible. And so the question of which population to sample from is an important one. Tests of local external validity are important.

ADJDM research can be folded into descriptive moral psychology simply by asking which ethical behaviors can be explained of the various different mechanisms postulated by its hypotheses—assuming these hypotheses are true. Indeed, models of moral cognition can be developed using convenience samples of WEIRD participants, and then assessed according to the degree to which they are able to explain broader patterns of ethical behavior described by hard-question social theory. As I noted earlier, this specific research question can be called the "container test" because it amounts to looking for evidence that a psychological model can be "contained" by a more general sociological theory. At the ideal limit, with the evidence suggesting that the relevant containment exists, the two theories will yield a multilevel theory of the structural and psychological processes that interact to produce a particular set of positive social outcomes. Conversely, complete failure of the container test is evidence that the relevant psychological model is unable to explain the relevant patterns of social behavior, and so the model's validity is restricted to the laboratory only.

One way of implementing these ideas would be to use causal graphs. If there is evidence that warrants inferring that an edge leads in from vertices that represent a stable property of a social institution to some nodes in a psychological model, and likewise, evidence that warrants inferring edges leading back out of the psychological model to vertices that are representations of other social processes, then the intuition is that the social theory "contains" the psychological theory. There are of course other formal and nonformal ways of representing what it would be to contain a psychological model within a social theory, and indeed, it would be helpful if examples useful for illustrating the container test were instances of existing psychological research. Unfortunately, I am not aware of any example of research in moral psychology that tries something like the container test. So the best I can do to further illustrate what the container test amounts to is to construct two hypothetical examples in which existing models of moral cognition are exported from the lab into hypothetical social theories.

The first example is a developmental theory of moral cognition proposed by Lene Arnett Jensen. It is called the *cultural-developmental template approach* (Jensen 2008; Jensen 2011a, b), and the template itself is provided below (figure 9.1). Jensen argues that the template represents how developmental time and cultural values interact in different populations to produce different moral perspectives.

There are two groups of analytical concepts in Jensen's model: there are developmental stages (childhood, adolescence, and adulthood) and

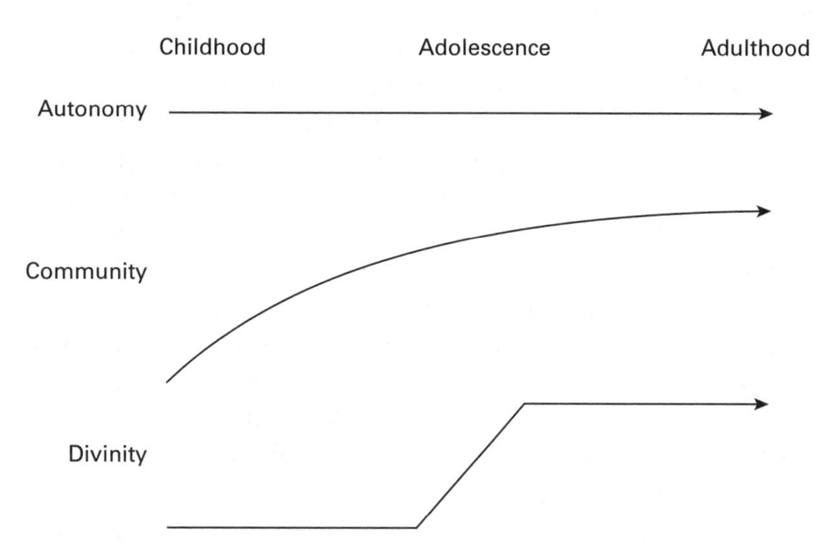

Figure 9.1
L. A. Jensen's cultural-developmental template of moral development.

three ethical orientations (autonomy, community, divinity). The second of these clusters is the "big three" moral domains taken from the past work of Richard Shweder, who with his collaborators argues that moral discourse can often be interpreted as being about one or another of these domains (Shweder et al. 1997; Shweder 2000). Jensen argues that different interactions between these concepts explain, for instance, religious liberals versus religious conservatives: by adulthood, liberals load high on autonomy and community and low on divinity, while religious conservatives load high on divinity and community, and comparatively lower on autonomy. Other normatively integrated populations can thus be defined by different developmental trajectories and outcomes.

How might Jensen's model be synthesized with an account of some existing social institutions? This cannot be done without some revision to the model, of course. For instance, Jensen's template only represents a single life span, and it does not automatically accommodate cross-generational processes or change. But more substantially, instead of interpreting the model as explaining either gross patterns in people's opinions or patterns in quasi-political constructs, we instead try to find a way of enclosing it within a particular social institution that is known to produce a particular set of good outcomes. Some intermediary categories may be necessary to link up the psychological with the level of description of the relevant institutions. Setting such details aside, and assuming for the sake of the example that we have a hard-question social theory that tells us that norms of friendship embedded in institutions of mass media are causally correlated with several other good outcomes in a study of some normatively integrated community, then figure 9.2 is how we might, hypothetically, reconfigure Jensen's model so that it is contained by said norms of friendship, at least for the relevant population.

Again, this is a completely fictional example, and so its accuracy or scientific plausibility is not at issue. The only relevant point is that this interpretation of Jensen's model shows how it can be situated within an existing normative framework, which therefore would be evidence of which social behaviors the model explains while also demonstrating that the model can explain at least some social behaviors. Likewise, if there were another array of social institutions that could not contain Jensen's model, this would be evidence that Jensen's model is not a general model of moral cognition and development.

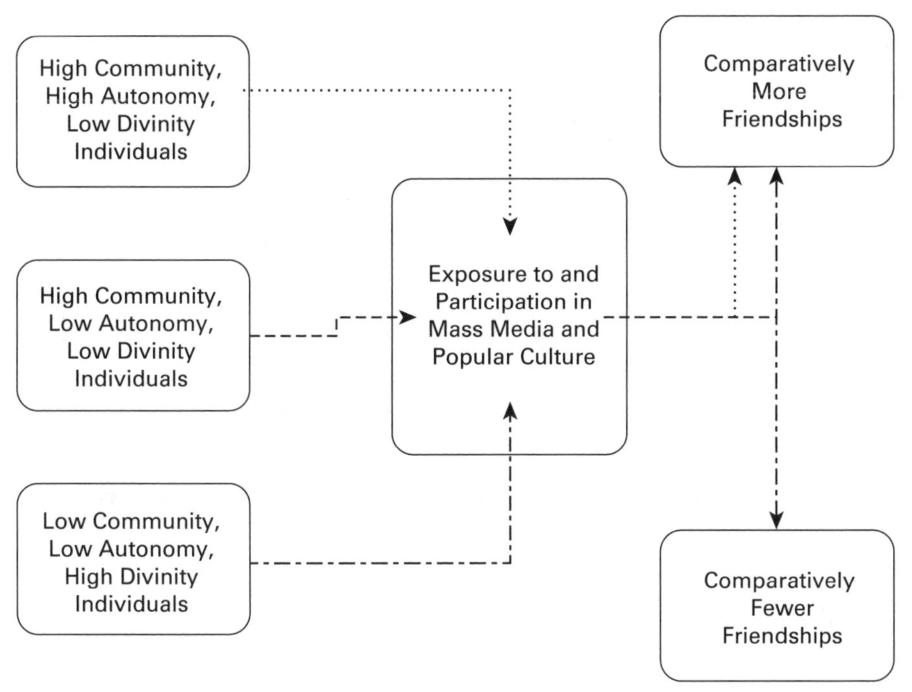

Figure 9.2

A hypothetical inference map illustrating how it is possible to reconfigure some of the analytical categories from Jensen's template so as to link these categories with the results of hard-question research in social theory.

We can use Haidt's social intuitionist model (SIM) as another example to illustrate what applications of the container test might look like. Haidt's model contains more information about the types of psychological processes that may constitute moral cognition, and the model makes some famously strong claims about how these processes can and cannot interact with one another. Haidt's conceptualization of the psychological content of the moral domain follows after Paul Rozin's earlier reinterpretation of Shweder's "big three" hypotheses, by proposing a model that grounds moral cognition not in three cultural "ethics" but in approximately five or six clusters of affective response (Haidt 2001). His model follows. Note that the picture of the model does not represent the clusters—which are harm/care, fairness/reciprocity, authority/respect, purity/sanctity, in-group/out-group (Haidt and Bjorklund 2007)—but rather the idea is that each cluster represents a dimension of the model that can explain different normative outcomes in a manner similar

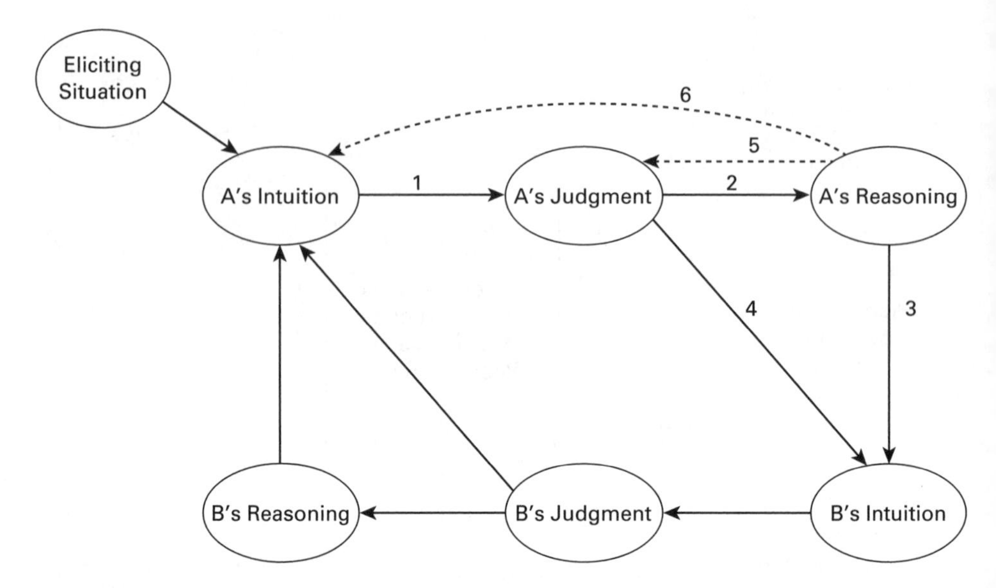

Figure 9.3
The social intuitionist model of moral judgment.

enough to Jensen's approach. According to the SIM, each of these clusters corresponds to a relatively innate and nonmalleable moral module, the output of which is the aforementioned affective responses, which Haidt somewhat confusingly calls "moral intuitions." For Haidt and his collaborators, one group of people may seek to remove from their cultural space stimuli the cause of an affective response in the purity/sanctity dimension, and otherwise amplify stimuli that trigger the other four modules, and in so doing create the psychological basis for a moral perspective organized around the latter four affective responses (figure 9.3).

One obvious way to see if the SIM is "contained" in any social institutions that produce ethical outcomes is to see if, for instance, institutions that embed norms for creating in-groups lead to some valuable outcomes. Indeed, Haidt has argued that his model is able to explain some of the fundamental socionormative processes and transformations of Western society, so there is some reason to think that the SIM should easily pass the container test and thus have its local external validity vindicated. Here, then, is a hypothetical example of how a group can be split as a result of institutional interactions with in-group ("us/them") intuitions (figure 9.4).

These two examples show how the container test provides a way of assessing whether descriptive models in moral psychology can explain

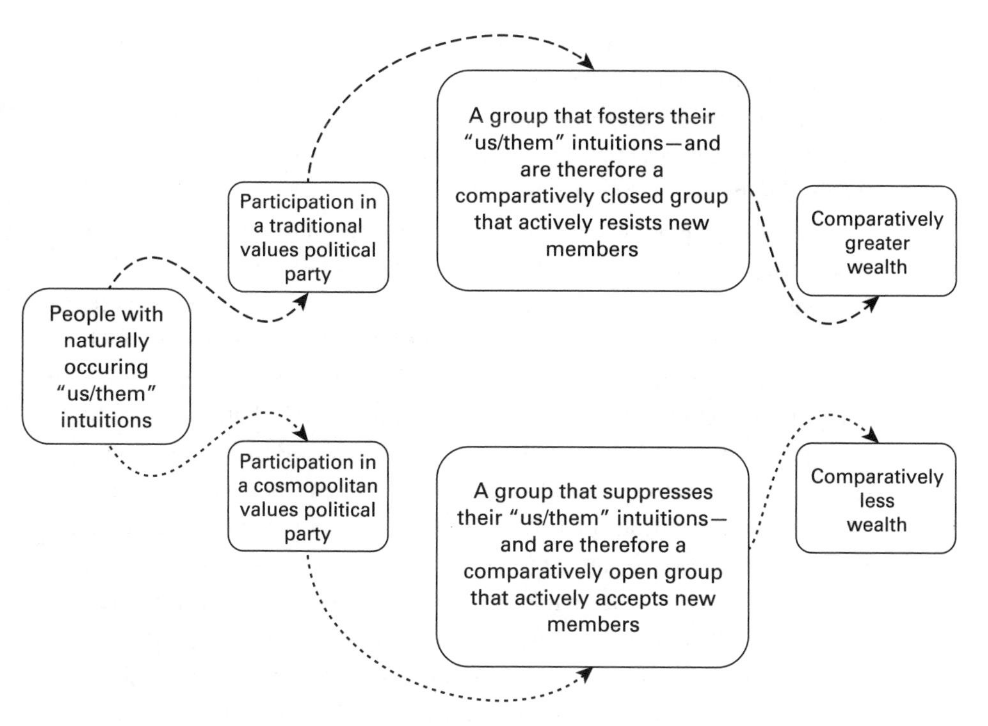

Figure 9.4
A hypothetical inference map illustrating how it is possible to reconfigure some of the analytical categories from the social intuitionist model so as to link these categories with the results of hard-question research in social theory.

social outcomes. When a model can pass the container test there is reason to infer that the psychological theory it represents can account for the psychological content and presuppositions associated with a social theory that provides an answer to a hard question—and therefore to infer that the model has local external validity at least. So, to put the same conclusion a bit more generally, one way of folding work in ADJDM moral psychology into descriptive moral psychology guided by hard-question social theory is to investigate whether or not patterns of ethical behavior can be explained by the relevant psychological hypotheses and models.

Microsystem Moral Psychology

As we have noted several times now, the concepts that are useful for describing how social institutions produce good outcomes will not also be apt for carrying information about the social and psychological phenomena that exist—to put the point perhaps too starkly—somewhere between national programs of public education and whatever low-level psychological processes regulate eye-gaze duration. How do we take into account the ethical thought and behavior that occurs in this middle space?

My proposal is to repurpose the justificatory principles embodied in the causal theory of ethics for the purpose of defining a conceptual vocabulary necessary to describe ethical behaviors that occur within this space, and then construe research into the cognition and motivation of which small-scale social behaviors are ethical according to these new principles as an important complement to the psychological research that, precedingly, is characterized as leading to container tests. Those are the philosophical ingredients for what can be called the "microsystem moral psychology," and the central question for this kind of moral psychological research is similar to that asked for the container test: What patterns of thought and feeling motivate ethical behaviors that occur at the level of family, friends, colleagues, and fellow church goers—the microsystem? As before, the guiding philosophical principle is that, if some kind of behavior causes a good outcome, then the behavior in question is ethical, and so understanding the psychological foundations of the behavior becomes a central question for descriptive moral psychology. Microsystem moral psychology, then, is the investigation of how small-scale good social outcomes are realized in patterns of social behavior that are small enough to be the subject of psychological investigation, and therefore are unlikely to be described by any independent hard-question social theory.

Another way to understand the idea behind microsystem moral psychology is to consider the following objection to the causal theory's tactical nominalism about the good and valuable. To wit: is not the good and valuable *too variable* and *too dependent* on dissimilar individual preferences and idiosyncratic social practices to have any objective reality? Indeed, we can even assume that further surveys of good and valuable outcomes provide conclusive evidence that, in certain large normatively integrated populations, the clusters of good and valuable that are realized have their own internal causal unity. The objection to this is that it would not follow from even that evidence that the good and valuable is real, because the relevant evidence would omit measures that are sensitive to the full range of all of the things that are both valued by people and realized by different patterns of human behavior. Once this full range of considerations is taken into account, any realism about the good and valuable loses its credibility because there will no longer be evidence of some kind of underlying causal unity. Said differently, the point is this: people value certain society-wide social goods (fair courts, free education, progressive taxation, stable prices for staple goods), but they also value certain outcomes that are by-products of small-scale social interactions (practicing yoga together, worshiping with friends of similar faith, the co-op housing program organized by a neighborhood), and they also value certain outcomes that are intrinsically personal (*my* physical health and strength, *my* skill at debate, *my* creative expression in dance). The objection, then, is that once these additional "nonstructural" good and valuable outcomes are added to the picture, the plausibility of the idea that *the* good and valuable could be real disappears.

But there is a compelling response to this objection. Metaphysically, there is a facilitation relation that holds between society-wide social goods and both different small-scale social goods and personal goods, and this facilitation relation is not transitive. Society-wide growth in health and wealth is valuable partly because, together, they serve as a platform that facilitates the realization of many different small-scale social goods and personal goods—but even aggregated together, these different small-scale social goods and personal goods do not cause the society-wide good outcomes. *My* enjoyment of cooking classes, and *our* local teen fiction novel reading club can be realized only because of the effects of a host of society-wide social institutions, but it does not cause these institutions. So, there is a metaphysical difference between the good and valuable outcomes that may constitute *the* good and valuable for some large normatively integrated populations and the many other kinds of (smaller, more prosaic) good and valuable things that small communities and individual people also value, and sometimes realize by their behavior. That difference is a reason to exclude the smaller-scale social and personal goods from the scope of our construct for *the* good and valuable.

It is very likely that the kind of psychological research that will be most illuminating of the small-scale social (and individual—but more on this in the next chapter) differences is research that integrates developmental, social, and physiological perspectives. For example, Felix Warneken writes:

> Evidence for helping early in ontogeny indicates that it is implausible that helping requires an adult-like moral value system, as preverbal infants are unlikely to be

motivated by normative principles. In fact, it is not until middle to late childhood that children begin to reason about social norms as obligatory. Only then do children begin to perceive failures to follow such norms as guilt-evoking, develop a moral self, and hold themselves and others to the same standards. Hence, young children may have a predisposition for altruistic behavior that is not based upon socialization factors such as reputational concerns, a long history of rewarding, or a rich moral value system alone. (Warneken 2016)

The more general methodological idea that can be extrapolated from Warneken's argument is that close study of the effects of socialization over developmental time may yield psychological hypotheses that explain at least some aspects of the social behaviors involved in the realization of small-scale social goods. For example, if children have a predisposition for altruistic behavior, then studies of how this disposition is shaped by culture may yield insight into the motivational and cognitive structures driving social behavior like helping. For an example, we turn to Warneken again: "Adults in one cultural context may view their children's helping as a social obligation, whereas in another they may construe helping as an expression of individual preferences."

And then, in a related study, Yarrow Dunham, Andrew Scott Baron, and Susan Carey report:

Children were randomly assigned to groups and engaged in tasks involving judgments of unfamiliar in-group or out-group children. Despite an absence of information regarding the relative status of groups or any competitive context, in-group preferences were observed on explicit and implicit measures of attitude and resource allocation (Experiment 1), behavioral attribution, and expectations of reciprocity, with preferences persisting when groups were not described via a noun label (Experiment 2). In addition, children systematically distorted incoming information by preferentially encoding positive information about in-group members (Experiment 3). (Dunham, Baron, and Carey 2011)

Dunham and coauthors speculate that these results can be explained by a "general affective positivity" that is created by being included in a social group. Perhaps this mechanism is linked with—and perhaps is even the same as—the predisposition to altruism identified by Warneken.

Note that Dunham and colleagues say that they deliberately avoided introducing any explicit social context manipulations their study. That is in contrast with an earlier study in which Megan Patterson and Rebecca Bigler examined the effects of "top-down" recognition of groups on children's attitudes about their peers:

> Children attending day care were given measures of classification skill and self-esteem and assigned to membership in a novel ("red" or "blue") social group. In experimental classrooms, teachers used the color groups to label children and organize the classroom. In control classrooms, teachers ignored the color groups. After 3 weeks, children completed multiple measures of intergroup attitudes. Results indicated that children in both types of classrooms developed in-group-biased attitudes. As expected, children in experimental classrooms showed greater in-group bias on some measures than children in control classrooms. (Patterson and Bigler 2006)

Speculatively, of course, if we begin to link the insights of these studies together, we can form an impression of the different developmental and social processes that interact to produce that kind of small-scale social groups that, in time, are the source of small-scale good outcomes. A Victorian novel reading group is, after all, but one kind of in-group among thousands and thousands of others.

So, the ultimate goal of descriptive moral psychology can therefore be quite ambitious: an integration of theories in physiology, developmental psychology, social and cognitive psychology, and social theory that tells a psychological (more precisely: proximate) story about how and why we pursue many of the socially realized valuable outcomes that are possible to experience. Container tests—or similar methodological ideas—seek to understand the connection between sociological descriptions of ethical behavior and cognitive psychology. Microsystem moral psychology complements that focus by drawing attention to the moral significance of interactions between developmental and physiological pathways and microsystem-level social processes.

Finally, a few words about a benefit of adopting the guidelines of descriptive moral psychology. Almost all researchers currently working in moral psychology have tacitly adopted a definition of ethical behavior that is psychological in nature. What makes some pattern of behavior *ethical behavior* is, specifically, that it is caused by some pattern of distinctly moral cognition. However, because they require researchers in moral psychology to reject the proposition that ethical behavior must be defined psychologically, there are a number of testable hypotheses that have not yet received much attention from researchers in moral psychology. These are specifically the hypotheses that cluster around the question of whether or not certain patterns of ethical behavior are caused by cognitive patterns and mechanisms that have been identified by independent research in (nonmoral) cognitive, social, and developmental psychology. One of the scientifically attractive features of adopting the theoretical package formed out of the causal theory of ethics and the guidelines that constitute descriptive moral psychology is that this package allows us to identify and investigate and try to resolve an empirical hole in contemporary moral psychology: we have the theoretical resources now that tell us how to investigate whether some independently identified patterns of nonmoral cognition are able to explain patterns of both moral and nonmoral behavior.

Conclusion

The causal theory of ethics gives descriptive moral psychology a principled reason to treat those social behaviors that produce some good outcome as ethical behaviors—and leads in turn to the methodological conclusion that the scope of descriptive moral psychology may be very, very large. However, it is important to be reminded again of the key implication of conceptual framework pluralism: it is not at all bad if multiple different scientific theories, models, and technical conceptual vocabularies are needed in order to yield a comprehensive understanding of normative behavior, which is after all something of rather marvelous complexity and diversity. In fact, an increase in the proliferation of different theories and models in moral psychology may be an important bellwether of scientific progress, but only so long as each of these theories does not strive for both perfect accuracy and universal external validity at the same time. There is no reason for a scientist or philosopher who is friendly to the idea of conceptual framework pluralism to think that moral psychology must be required to use only a single unified conceptual vocabulary to explain *all* patterns of ethical thought and behavior.

It is an implication of this argument that many existing research projects in fields of human development, social psychology, and education and related areas have elements that may count as examples descriptive moral psychology. This should not be as much of a surprise as it may first seem. Many behavioral scientists are drawn to trying to understand not just the human organism as it is found, but also, following in the vein of Foot (2001), what is good for the organism that we are. Descriptive moral psychology is likely therefore to be an interdisciplinary project that runs through many otherwise methodologically autonomous fields of psychological and behavioral science. The picture of ethical knowledge that fits this interdisciplinary project is exactly the one that is implied by the causal theory of ethics: it is a picture that does not imply that ethical knowledge must add up to, or really make sense, in terms of some overarching universal, transhistorical, transcultural normative theory. In other words, the picture of ethical knowledge that fits this conception of moral psychology is one that makes ethical knowledge out to be very much like knowledge of social behavior in the biological world: the knowledge is a collection of insights into local behavioral regularities and their effects, and perfectly general principles from which these insights can be derived are unlikely to exist.

That is a view of ethical knowledge that is not common in Western moral philosophy. Indeed, it is customary for contemporary moral philosophers to permit themselves a great deal of abstraction, so that debates about environmental ethics or just-war theory have almost no inferential connections with existing historical facts or policy option (Heath 2015). There is also the risk that a similar state of affairs will be realized in moral psychology, albeit by different intellectual mechanisms than those that drive philosophers to ever higher levels of abstraction and empirical disconnect. The mechanism in psychology is what Helen Longino sometimes calls decontextualization, where, for the purposes of collecting clean data or permitting straightforward statistical analyses, complex social behaviors like aggression are studied by searching for a proxy behavior that is assumed a priori to have all of the essential features of aggression but, because they occur in highly constrained, easy-to-observe environments, permit themselves to be studied more easily. The question, then, is whether the proxy behavior—like prison violence, or "microaggressions" elicited by a subtle stimuli—are sufficiently like the target (nonproxy) behaviors (Longino 2013). And this is not an abstract question. The social or moral importance of the target behavior can determine how much, if any, decontextualization is appropriate. As we have noted, ADJDM moral psychology is committed to an extremely decontextualized approach to moral behavior, and both the container test and microsystem moral psychology provide ways of de-decontextualizing existing theories in moral psychology. The argument for pursuing this de-decontextualization is simply one of minimizing risk: when the authority and prestige of science are used to ratify conclusions about human moral nature, it matters very much that the science be as rigorous as possible.

Indeed, it has not yet been widely appreciated that there is a fairly serious downside risk to trying to study morality both scientifically and in a decontextualized fashion. Nietzsche was probably right that no intrinsic facts about human psychology prevent us from moralizing about any norms or behaviors we choose, and this means that we face a question of which norms, if any, we should in fact moralize (for further development of this idea, see Williams 1993). The suggestion advanced so far is that evidence about the population-level effects of adopting certain norms can be used to make decisions about which norms to moralize. But an alternative picture holds that a combination of ADJDM moral psychology and research into the evolution of morality refutes the Nietzschean idea that psychology places no deep constraints on what we moralize. According to this non-Nietzsche alternative, we know from evolutionary theory that the mind will inherently moralize certain patterns of behavior, because in the environment of evolutionary adaptedness (EEA) there were fitness benefits to pursuing or resisting the relevant behaviors, and so instincts for these behaviors were selected for. And we know from ADJDM moral psychology's acceptance of the reality of moral dumbfounding that the insights of philosophical ethics are seen as irrelevant to how people think and act, ethically speaking. The confluence of these two lines of reasoning implies a kind of quietism about the inevitability morality according to which there is not much we can do to escape its influence, as per Barash (1979), nor much we can do to rationally control its subsequent direction. If scientific research in moral psychology seems to be painting a picture according to which we cannot rationally choose which norms to moralize, then there is the risk that we choose not exercise this control. That, as I say, is a rather important downside risk to take into account. Descriptive moral psychology is a way of taking account of this risk, as it consists of different methods that can either assess whether a decontextualized theory in moral psychology has gone wrong, or otherwise ensure that theories in moral psychology remain appropriately grounded in facts about existing social practices and behaviors.

Still, let me conclude this chapter on a more positive note. I have tried to describe both the container test and microsystem moral psychology as clearly and concretely as possible. But the reality, both conceptually and methodologically, is that there will not be a logical gap separating ethical inquiry and descriptive moral psychology. This is because there can be—and indeed, already are—various kinds of feedback loops running between efforts to form or modify existing social institutions at both the large and small scale and the various kinds of psychological research that can contribute moral psychology. Far from debunking the rationality of ethics, the implication of the existence of these feedback loops is that scientific moral psychology can make a genuine contribution to both epistemic and metaphysical progress in ethics. So long as ethics is conceived of naturalistically, which is of course how the causal theory of ethics conceives of it, moral psychology can do rather a lot more than attempt to debunk ethics (Kumar and Campbell 2012). Moral psychology can in various ways make extremely important contributions to epistemic, and perhaps even metaphysical, progress in ethics. This of course requires finding a way of understanding the relationship between moral psychology and normative ethics that avoids holding either that psychological science is irrelevant to normative ethics (Berker 2009) or that normative ethics can be eliminated, in either whole or part, by moral psychology (Greene 2014). And at this point, we have most of the pieces of just such an understanding laid out. All that remains is our discussion of normative moral psychology.

Thus, by setting as one of the criteria governing the acceptance of theories in moral psychology the requirement that any theories or models explain at least some observed social behaviors that are known on independent grounds to be connected with valuable outcomes, we ensure that all accepted theories in moral psychology will have at least local external validity and also that these theories will not generate inconsilience between moral psychology and philosophical ethics. The downside to adopting such criteria is that container tests are, quite literally, costly. These tests can only be formulated on the basis of an interdisciplinary understanding that combines expert knowledge from the behavioral and social sciences. Furthermore, the data collection protocols for the tests will often be more complex, and thus more costly to design and run, than sampling the hypothetical judgments of undergraduates. But all the same, we now have some reasons to think these are costs we should be willing to pay.

References

Barash, D. P. 1981. *The Whisperings Within: Evolution and the Origin of Human Nature*. New York: Penguin Books.

Berker, Selim. 2009. "The Normative Insignificance of Neuroscience." *Philosophy & Public Affairs* 37 (4): 293–329.

Bronfenbrenner, Urie. 1979. *The Ecology of Human Development.* Cambridge, MA: Harvard University Press.

Dunham, Yarrow, Andrew Scott Baron, and Susan Carey. 2011. "Consequences of 'Minimal' Group Affiliations in Children." *Child Development* 82 (3): 793–811.

Foot, P. 2001. *Natural Goodness*. Oxford: Clarendon Press.

Greene, Joshua D. 2014. "Beyond Point-and-Shoot Morality: Why Cognitive (Neuro) Science Matters for Ethics." *Ethics* 124 (4): 695–726.

Gurin, P., E. L. Dey, S. Hurtado, and G. Gurin. 2002. "Diversity and Higher Education: Theory and Impact on Educational Outcomes." *Harvard Educational Review* 72 (3).

Haidt, J. 2001. "The Emotional Dog and Its Rational Tail: A Social Intuitionist Approach to Moral Judgment." *Psychological Review* 108 (4): 814–834.

Haidt, J., and F. Bjorklund. 2007. "Social Intuitionists Answer Six Questions about Moral Psychology." In *Moral Psychology: The Cognitive Science of Morality*, vol. 2, ed. Walter Sinnott-Armstrong, 181–217. Cambridge, MA: MIT Press.

Heath, Joseph. 2015. "Kymlicka on Interculturalism vs. Multiculturalism." *In Due Course*, March 18. http://induecourse.ca/kymlicka-on-interculturalism-vs-multiculturalism/ (accessed March 20, 2015).

Henrich, Joseph, Steven J. Heine, and Ara Norenzayan. 2010. "The Weirdest People in the World?" *Behavioral and Brain Sciences* 33 (2–3): 61–83, discussion 83–135.

Jensen, Lene Arnett. 2008. "Through Two Lenses: A Cultural–Developmental Approach to Moral Psychology." *Developmental Review: DR* 28 (3): 289–315.

Jensen, Lene Arnett. 2011a. "The Cultural-Developmental Theory of Moral Psychology: A New Synthesis." In *Bridging Cultural and Developmental Psychology: New Syntheses in Theory, Research and Policy*, ed. L. A. Jensen, 3–25. New York: Oxford University Press.

Jensen, Lene Arnett. 2011b. "The Cultural Development of Three Fundamental Moral Ethics: Autonomy, Community, and Divinity." *Zygon®* 46 (1): 150–167.

Kumar, Victor, and Richmond Campbell. 2012. "On the Normative Significance of Experimental Moral Psychology." *Philosophical Psychology* 25 (3): 311–330.

Longino, Helen E. 2013. *Studying Human Behavior: How Scientists Investigate Aggression and Sexuality*. Chicago: University of Chicago Press.

Pärnamets, Philip, Petter Johansson, Lars Hall, Christian Balkenius, Michael J. Spivey, and Daniel C. Richardson. 2015. "Biasing Moral Decisions by Exploiting the

Dynamics of Eye Gaze." *Proceedings of the National Academy of Sciences of the United States of America* 112 (13): 4170–4175.

Patterson, Meagan M., and Rebecca S. Bigler. 2006. "Preschool Children's Attention to Environmental Messages about Groups: Social Categorization and the Origins of Intergroup Bias." *Child Development* 77 (4): 847–860.

Phillips, Jonathan, and Alex Shaw. 2015. "Manipulating Morality: Third-Party Intentions Alter Moral Judgments by Changing Causal Reasoning." *Cognitive Science* 39 (6): 1320–1347.

Rai, Tage Shakti, and Alan Page Fiske. 2011. "Moral Psychology Is Relationship Regulation: Moral Motives for Unity, Hierarchy, Equality, and Proportionality." *Psychological Review* 118 (1): 57–75.

Shweder, R. A. 2000. "Moral Maps, 'First World' Conceits, and the New Evangelists." In *Culture Matters: How Values Shape Human Progress*, ed. Lawrence E. Harrison and Samuel P. Huntington, 158–172. New York: Basic Books.

Shweder, R. A., N. C. Much, M. Mahapatra, and L. Park. 1997. "The "Big Three" of Morality (Autonomy, Community, Divinity) and the "Big Three" Explanations of Suffering." In *Morality and Health*, 1st ed., ed. Allan M. Brandt and Paul Rozin, 119–169. New York: Routledge.

Warneken, Felix. 2016. "Insights into the Biological Foundation of Human Altruistic Sentiments." *Current Opinion in Psychology* 7 (February): 51–56.

Williams, Bernard. 1993. "Nietzsche's Minimalist Moral Psychology." *European Journal for Philosophy of Science* 1 (1): 4–14.

10 Normative Moral Psychology

Enabling Knowledge and Functional Virtues

In chapter 3 I defined the distinction between the psychological representation of the structural content of a regulative framework, and a different kind of knowledge, *enabling knowledge*. Behind this distinction is the notion that there is more than one kind of normative success that is possible when playing games or, indeed, engaging in any kind of social behavior that involves the realization of regulative frameworks. The first kind of normative success is just what constitutes the realization of the rules: the rules or norms get put into practice. But there are many other kinds of success that can complement that first kind; these additional forms of success are all of the very many ways of "doing well" within a normative context. Concretely, this can emerge as the difference between knowing how to play a game of hockey and knowing how to win every game of hockey, or knowing how to play a game of hockey and also knowing how to score a goal on almost every attempt, or knowing how to play a game of hockey and also know how to enjoy the physical exercise and the pressures of competition. The first kind of success explains why a player is not constantly sent to the penalty box, or why the other team does not simply abandon the game because they perceive, rightly, that their opponents do not know the rules—that kind of success is caused by accurate psychological representations of the structural content of the relevant regulative framework forming the rules of hockey. But there is more to the game of hockey than just playing by the rules; and there is much more to doing well in a social world created by social institutions than just following the rules. Knowledge of how to complement the successful realization of a regulative framework with the realization of some other valuable end is *enabling knowledge*.

Enabling knowledge is a very abstract concept, so here is an alternative explanation for it. When we realize any set of norms, there is a sense in which our behavior is thereby correct; incorrect behavior results in failing to realize the relevant norms. But even when we follow the rules as correctly as possible, we can nevertheless do this efficiently or wastefully, competently or awkwardly, with patient kindness or with spiteful mendacity, and so on. This is because of the elementary metaphysical fact that any choice or action always has both multiple intended and unintended effects, and so it simply is not possible to *only* realize a regulative framework. Even consciously trying to do just that would result in someone being, for instance, exceedingly literal-minded about some particular social behavior, which would then amount to being peculiar. Enabling knowledge is the name for the knowledge an individual person can acquire that provides them with insight into how to align their role in the realization of different regulative frameworks with some additional set of valuable ends. Through enabling knowledge people realize all manner of desirable ends within different normative contexts, while at the same time engaging in "regulatively correct" behavior.

To return to the hockey example one last time, it is the enabling knowledge of each player on a team that explains why that team wins, or enjoys, every game it plays—but this is not to say that each player on the team has exactly the same enabling knowledge. Each player will have their own natural talents and abilities, own habits of play, own psychodynamic relationship with the game of hockey, and so on. While it is the team as a whole that enjoys a season of victory, for example, this pattern of outcomes is caused by each player's own unique and individual skill. That particular feature of enabling knowledge is what, in conjunction with the metaphysical fact that all actions always have multiple effects, allows enabling knowledge to be a source of the realization of deeply personal ends. Enabling knowledge, thus, can benefit both any person who has it and the other people with whom they are cooperating; actions caused or facilitated by enabling knowledge can be neither entirely selfish nor wholly prosocial.

At this point I introduce one final technical concept to our discussion. This is the idea of a *functional virtue*. The definition is this:

A functional virtue is enabling knowledge that is positively associated with the realization of a pattern of ethical behavior, where "ethical behavior" is defined according to the causal theory of ethics, and where the relevant enabling knowledge also helps to, but need not always, cause the realization something of personal value to the individual with the knowledge.

So, if games of hockey were direct causes of world peace, then a player's individual excellence at hockey would be a functional virtue if, for instance, the player also deeply valued their facility for the game. Slightly more realistically, since the social institutions that create public schools are typically a cause of various good social outcomes, being an effective teacher is (or would be) a functional virtue if all public school teachers also, as individuals, value their own happiness and being an effective teacher contributes something meaningful to each person's happiness. Put more generally: functional virtues are the enabling knowledge that allows a person to align the realization of at least some of their own personal values with the patterns of cooperation that are the cause of good and valuable social outcomes. Functional virtues are not necessary for the social institutions to be formed which cause realizations of the good and valuable; but their compatibility with the other psychological components of the foundations of different social institutions can explain why the social institution is stable, or readily attracts new participants, or is able to share a normative context with a variety of other social institutions. That is not to argue that the existence of functional virtues provides self-interested people with a universally compelling reason to be ethical, or if not that, then at least to be prosocial. Rather, their existence shows that it is possible to align self-regard with regard for others in ways that are mutually sustaining and perhaps even self-reinforcing.

Is there evidence that functional virtues really exist? Yes, and some examples that they do are given in the remainder of the chapter. The proposal here is that normative moral psychology is the search for functional virtues, and therefore the fact that some functional virtues have already been discovered is evidence of the viability of normative moral psychology. Accordingly, we can define "normative moral psychology" as research into the cognitive mechanisms, habits of body or mind, and other characterological traits that are able to link the sphere of personal value and self-regard with those cooperative actions that produce the kinds of social goods that can be described by hard-question social theory.

Using the idea of functional virtues, we can distinguish between *moral* and *ethical* in a principled fashion, by restricting the moral domain to the world of a developed self with personal values, psychological abilities, and associated characterological traits, and setting the ethical domain to be the world of social institutions, agent-neutral social roles, large-scale intergenerational cooperation, obligatory norms that transcend personal relations, and complex good and valuable social outcomes. The key to this distinction is to see that a person can be ethical without being moral, simply by participating in some social behaviors that are ethical and yet being motivated to do so by something other than any functional virtue. However, being moral requires being ethical: being moral just is to have the practical knowledge that allows a person to increasingly improve their ability to unify practicing self-regard for their own values with participation in the different forms of ethical (and therefore inherently social) behavior. So, an alternative definition of normative moral psychology is that it is the study of what it is to be moral in the social contexts where ethical action is really possible. Descriptive moral psychology (as defined in chapter 9) is, therefore, about what it is to be ethical without necessarily being moral.

Just as importantly, the distinction between moral and ethical is desirable from the perspective of conceptual framework pluralism. A pluralist sees a strategic advantage for any scientific or philosophical project that is able to regiment families of concepts sufficiently well so that different inferential borders and conceptual "centers of gravity" appear. And so here the proposal is that normative moral psychology has as its "center of gravity" mostly characterological phenomena while descriptive moral psychology has as its center the psychological foundations of small-scale prosocial and larger-scale ethical behavior.

Another way of getting at the distinction between moral and ethical is to think about young children, who at around the age of four are capable of following (or imitating, at least) many norm-regulated behaviors. In certain cases their behavior could be ethical, and so there may be a subfield of descriptive moral psychology that studies the psychological basis of normatively appropriate behavior in children. But there is unlikely to be a corresponding subfield of normative moral psychology—because children lack fully developed senses of self and of personal value. Absent such knowledge, a child can be at most "pre-moral"; becoming moral requires that children learn to construct an increasingly richer and more sophisticated sense of self and value, until it becomes possible to align their personal values with different ethically valuable social ends. Further empirical research into children's developing a pre-moral sense of self and value may therefore be crucial input to normative moral psychology.

Empirical Evidence

So much for the philosophy of normative moral psychology. Is there empirical evidence that normative moral psychology is a viable intellectual project? Is there reason to think that psychologists can discover functional virtues?

To answer those questions, we need examples of scientific research that show that people can learn how to better realize in their personal lives things that they value, where this success likely contributes to the ability to engage any number of social behaviors that are, at least probably, ethical in nature. But since no behavioral scientist has actually set out with the intention of studying functional virtue as such, some reading between the lines is inevitable. The key to justifying such readings is to look for research in the behavioral sciences that is able to answer yes to the following two questions: Does the research describe a developmental or cognitive process according to which people learn how to better realize something that they personally value? And is it likely that hard-question social science would confirm that people who have learned how to produce something of personal value would be, for that, more likely to participate in patterns of ethical behavior? Here now are four examples of research programs that answer these questions in the affirmative.

Happiness

Happiness is very well studied in psychology (Kahneman, Diener, and Schwarz 1999), and this research provides both strong evidence that happiness is an important personal good, and that it often has distinctly characterological causes. Michael Argyle gives a list of causes and correlates of happiness (Argyle 1999), and he reports that, historically, efforts to explain happiness as a by-product of structural (mainly demographic) factors account for less than 15 percent of the reported data. So while there are surely structural causes linked to a person's happiness, these are hardly the whole story. The rest is filled in, it seems, by any number of characterological traits in a complex network (for examples, see Lu 1999; Shimai et al. 2006; Park and Peterson 2006). These characterological traits—for example, openness to giving and receiving love, hopefulness, and energy—are, arguably, functional virtues.

Furthermore, there are also efforts being made to better organize and relate empirically contentful constructs for happiness with constructs for subjective well-being (Diener et al. 1999; Diener 2000; Samman 2007), and also to develop cross-cultural measures of the latter that are reliable (Samman 2007). This work is important for two reasons. First, it highlights the empirical possibility that functional virtues may be interdependent, and so further tests of the interdependence of functional virtues represent an intriguing line of new empirical research for normative moral psychology. Second, deeper and more systematic scientific study of the cultural variability of functional virtues may pay moral dividends, insofar as furthering this research may uncover different patterns of learning and socialization that are effective sources of functional virtues, allowing, in a sense, for moral cross-fertilization.

Mindset

Carol Dweck's distinction between "fixed" and "growth" mindsets is famous (Dweck 2000; Elliot and Dweck 2007, chap. 8), and her concepts are relevant here because they both denote characterological traits, and the growth mindset has been shown on the basis of very detailed experiments to be connected with several positive outcomes. A growth mindset is therefore a paradigmatic functional virtue.

People with a "fixed" mindset tend to assume that intelligence is a static entity, and they tend to explain success in terms of intelligence and failure in terms of the absence of intelligence. People with a "growth" mindset instead think of intellectual competence as something that can grow and develop, and thus tend to explain success and failure in terms of effort and motivation. People with growth mindsets are said to enjoy challenges, whether or not they succeed; and people with fixed mindsets find that failing a challenge sometimes damages their sense of confidence and sense of self-worth. Some of the positive outcomes associated with a growth mindset are positive moods and increased resilience in the face of a challenge, and it is hardly speculation to infer that these character traits can in turn be connected with increased ability to experience many other deeper and more enduring good and valuable personal outcomes.

More recent work has explored the effects of growth mindset interventions in at larger scales than those that are possible to create in the context of psychological experiments. In a recent study of about fifteen hundred high school students, David Paunesku and colleagues found the following: "Among students at risk of dropping out of high school (one third of the sample), each [growth mindset and sense-of-purpose] intervention raised students' semester grade point averages in core academic courses and increased the rate at which students performed satisfactorily in core courses by 6.4 percentage points" (Paunesku et al. 2015). Since education is a good created by social institutions, there are quite likely any number of norms that are ethical by the lights of the causal theory of ethics across different education programs. What this intervention does, then, is foster a functional virtue that is associated with these norms.

Positive Sentimental Order

A different approach to looking for functional virtues takes as its starting point a concept introduced by Anselm Strauss: *sentimental order*. In its original use, the concept refers to different patterns of mood and sentiment that influence the behavior of the workers and patients in an organization like a hospital. Strauss writes:

> A disruption of the ward's organization of work is paralleled by a shattering of its characteristic 'sentimental order'—the intangible but very real patterning of mood and sentiment that characteristically exists on each ward. For instance in an intensive care unit where cardiac patients die frequently, the sentimental order is relatively unaffected by one more speedy death; but if a hopeless patient lingers on and on, or if his wife, perhaps, refuses to accept his dying and causes "scenes," then both sentimental and work orders are profoundly affected. ... On obstetrics services, the characteristic sentimental order is one of relative cheer and optimism. When a delivering mother unexpectedly dies, as occasionally happens, the characteristic chaos that follows is a visible sign of the shattered sentimental pattern. (Strauss 1968, 14)

Strauss goes on to analyze how different sentimental orders facilitate the functioning of different kinds of hospital care. And for our purposes, the idea here is that, in its psychological embodiment, positive sentimental order is a—very prosaic—functional virtue. Learning and internalizing the moods and feelings that constitute the good sentimental order of a clinical, or a classroom, can be a kind of epistemic achievement that, in turn, helps sustain the creation of whatever goods can be produced in the course of everyday work, some of which may be truly ethical outcomes. In the case of a clinic specifically, this might be a reduction in the overall level of stress among clinicians and patients, and in the classroom, this may involve lectures that are less boring and more entertaining for everyone involved.

Purpose

Purpose is beginning to emerge as an important construct in psychology. Erik Erikson defined it as "the courage to envisage and pursue valued goals uninhibited by the defeat of infantile fantasies, by guilt and by the foiling fear of punishment" (Erikson 1994, 122). More recently, Patrick McKnight and Todd Kashdan define purpose as "a central, self-organizing life aim that organizes and stimulates goals, manages behaviors, and provides a sense of meaning" (Hill and Burrow 2012). Of particular interest are the positive benefits of developing a sense of purpose: these include reduced anxiety, increased sense of subjective well-being, and, crucially for the connection with ethical behavior, increased comfort with ethnic diversity (Burrow et al. 2014; Burrow, O'Dell, and Hill 2010). A sense of purpose is therefore most likely another important functional virtue.

Objections, Responses, and Other Methodological Matters

A working understanding of functional virtues makes it easy to read through the existing literatures in developmental and social psychology and find further examples of normative moral psychology. But this should not be a surprise. It is very hard to formulate psychological hypotheses that describe causes of social behavior that are both explicitly and implicitly neutral with respect to value, simply because our thoughts and motivations are deeply connected to the outcomes we value (Rödl 2007). And it is an intriguing empirical idea that we may discover yet further ways of aligning ethical and moral thought and behavior.

It is now time to turn to a defense of the ideas behind descriptive and normative moral psychology, as well as some of the arguments from preceding chapters. I begin with the objections most closely related to normative moral psychology.

But Hasn't Psychology Shown That Virtues Don't Exist?

The first of these objections can be put simply: Hasn't social psychology shown that the virtues do not exist? Is it not a mistake to posit the existence of functional virtues? John Doris puts the problem this way:

> Aristotelian virtue ethics, when construed as invoking a generally applicable descriptive psychology, may appear more attractive than competitors such as Kantianism and consequentialism, in that characterological moral psychology might allow a more compelling account of moral development and agency. But understood this way, character-based approaches are subject to damaging empirical criticism. (Doris 1998, 520)

The empirical criticism is that "characterological moral thought may … foster a dangerous neglect of situational influences" (Doris 1998, 519). This is because "social psychologists have repeatedly found that the difference between good conduct and bad appears to reside in the situation more than in the person" (Maria, John, and Harman 2010, 357). For example, studies in the 1970s showed that some people are significantly more likely to help others who are in mild distress after they have found a trivial amount of money (Doris and Stich 2006). The argument, then, is not that there are no characterological traits, but rather that reference to characterological traits makes for much weaker explanations of why people do good and bad things than reference to situational factors. Situational influences are an explanatory gap for the theory of virtue, suggesting that the theory should not be used to explain good and bad action.

There are several responses to this objection. For example, a relatively straightforward reply would be to point out that the functional virtues are immune to this criticism. It is true that functional virtues are characterological traits; but the construct is not intended to help explain *all* facts concerning a person's moral development and agency, and it certainly is not the case that right, good, or appropriate action must defined in a conceptual framework that includes the concept of a functional virtue. Indeed, the causal theory of ethics shows that it is possible to define good and bad—that is: or ethical and unethical—action without recourse to any psychological concepts whatsoever. The position developed in this book is that the full story of moral development, agency, and appropriate moral decision making can only be understood by a plurality of conceptually autonomous theories generated by different social and behavioral sciences, as well as philosophical ethics. It is this grand package of theory, data, norms, and different kinds of analysis that must show us how to reconcile a commitment to the existence of morally worthwhile character traits and the situational influences on thought and behavior—not the theory of virtues specifically.

But further development of that idea presently. For now, there is a deeper and better response to the situationist objection that has already been given by Julia Annas. The paradigm case of situational influence for Doris and his collaborators are Milgram's experiments. However, recall from our earlier discussion Milgram's own account of the trauma suffered by participants in the original studies. This trauma helps us see that, as Annas writes, "What the subjects in the situationists' experiments found was that they had not developed the kind of practical intelligence that would have dealt appropriately with unforeseen situations and pressures; presumably they had not thought that they needed to put in this effort" (Annas 2005, 639). Since social psychologists, like most scientists, are perceived to be trustworthy, it is hardly a surprise that people are so easily walked into Milgram-style moral failings, and thereafter so deeply unsettled by what they have done. It is also no surprise to find out that when participants in studies involving deception in the form of subliminal priming are told they are being primed and so are conscious that they are being deceived, the effect of the priming quite easily disappears (Verwijmeren et al. 2013). Virtues, like most other complex cognitive abilities, are learned; and part of what leads to their acquisition is knowledge of one's own failures. Learning about those times where trying to do some small good, or even just avoid harm, one nevertheless causes some bad outcome can be a cause of increased virtue, functional or otherwise.

Haven't You Just Made Up a Concept to Fit the Psychological Literature?

It may also be objected that the concept of a functional virtue itself is ad hoc, because it has been invented simply to ground the argument that normative moral psychology already is alive and well.

The reply to this objection is to point to other contemporary accounts of virtue that have both functionalist elements and social dimensions. Nancy Snow's recent work is perhaps the best illustration of this recent trend. Snow defines virtues as a kind of social intelligence, where social intelligence is a "complex panoply of cognitive affective processes, and it helps people in social living" and the virtues "will not help us to advance any social goals, but only a subset of these" because "the motivations characteristic of virtues distinguish them from other forms of social intelligence" (Snow 2010, 85). Importantly, the definition of social intelligence that Snow's conception of virtue rests upon is itself developed as an extension of Walter Mischel and Yuichi Shoda's cognitive-affective personality system (Mischel and Shoda 1995; Mischel and Shoda 1999). Snow, too, finds it important to show that her conception of virtue answers to previously described constructs in the psychological literature.

Mark Alfano's work provides another example that helps overcome this objection. Alfano has recently proposed that the virtues be defined analytically. Part of the trouble with locating virtue among the known ontology of the human mind, is that, as Alfano puts it, "What's being asked [by the question, What is virtue?] is not simply whether people *are* virtuous, but whether they *can be* virtuous. It could turn out that even though no one is virtuous, it's possible for people to become virtuous. This would, however, be extremely surprising. Unlike other unrealized possibilities, virtue is almost universally sought after, so if it isn't widely actualized despite all that seeking, we have fairly strong evidence that it's not there to be had" (Alfano 2015).

Alfano defines the virtues by showing how a Ramsey sentence can be constructed so as to capture common-sense platitudes that may exist about what causal connections there may be among desires, courage, generosity, action, and so on, without also entailing that virtue is patently easy to acquire and coextensive with every plausibly moral deed.

But for our purposes it does not matter whether or not Snow or Alfano have exactly the right story to tell about virtues; their work shows that it is not ad hoc to introduce the notion of a functional virtue. And indeed: one of the benefits of grounding the whole theory formed by the causal theory of ethics and descriptive and normative moral psychology in the crucible of conceptual framework pluralism is that, should either Alfano's or Snow's or Aristotle's theory of virtue prove superior to other accounts of virtue, then it should be possible to replace the inferior theory with the superior theory—just so long as, *pace* Doris, it is understood that the theory of virtue is only one element in the larger, messy overall picture of human normative behavior, and therefore cannot and should not have among its epistemic aims the ability to answer all important moral or ethical questions that we can formulate.

That conclusion is also one of the reasons why a principled distinction between the constructs *moral* and *ethical* is useful to normative theory: the distinction provides us with way of keeping separate two categories of normative questions, thereby giving us a clearer sense of what, metaphysically speaking, we should and should not investigate for answers to those questions. The two categories are questions about human psychology and questions about the structure of human populations, and our metaphysical inventory of "things" to investigate for answers now includes, inter alia, functional virtues, other characterological traits that count as enabling knowledge, social institutions, potentially fundamental norms, the good and valuable, the psychological foundations of social institutions, and so on. The fundamental methodological principle of a pluralist approach to human normativity is mix-and-match until the best overall—and not necessarily internally coherent or completely consistent—theoretical package emerges.

Conclusion

One of the properties that makes the causal theory of ethics different than traditional philosophical theories in ethics is that it does not presuppose that the normative arrangements that produce ethical outcomes in local situations can be identified using a context-independent approach. Instead, evidence about the local effects of social institutions is what guides decisions about which behaviors to regard as ethical. The concept of functional virtues that provides the basis for normative moral psychology is very much an effort to remain true to the same underlying methodological idea. Functional virtues are, like the concept of an ethical norm in the causal theory of ethics, defined *functionally*: the concept provides researchers with a template that can guide investigations that can subsequently uncover evidence about which cognitive processes satisfy the definition. Or, it may turn out that the template is useless, and must be discarded; or its use may in time suggest an even better alternative. But in any case, the template is deeply compatible with there being many different ways of developing local functional virtues.

What is put forward, consequently, is a theory of the virtues that, like my earlier treatment of the good and valuable, depends upon a kind of tactical nominalism. The philosophical suggestion at work here is that the definitions in these theories will be revised, split, and made more complicated as a result of attempting to accommodate them better and better to the complexity of both human psychology and social behavior. What's more, since these are definitions of normative concepts, it really does matter that these definitions do not foreclose the possibility of moral progress by reducing, for example, functional virtues to some set of already existing intellectual abilities. But it is equally important that we use concepts that are not so abstract or idealized that the only way we can imagine the world providing these concepts with empirical content is for the world to become, finally, morally perfect. It is one thing to define concepts that can be used in theories that describe normatively ideal circumstances, and then suggest that these theories serve as regulative ideals. But it is quite another thing to try to formulate concepts that are useful for the empirical work that needs to be done to get from, well, the moral world we actually inhabit to what may be only a slightly better world. For this work, our concepts need both the flexibility and the substance to find ways of accommodating themselves to existing causal structures.

Indeed, universal theories of either ethics or virtue risk seeing the failure of the world to conform to definitions in these theories as, for instance, the failure of the world to live up to its normative potential. Differences between theory and practice are therefore easily treated as evidence that the world is not yet normatively optimal, and that it is largely the world's responsibility to discover how to bend itself toward the pure vision of the theory. The point of these remarks is that any theory of normative thought and action needs for itself a theory of reference and truth that allows normative concepts to describe a dynamic and changing social reality, instead of either the world just as it is found or the world just as it would be if all our normative ideals were realized. I suggest that the notion of mutual accommodation of different conceptual vocabularies to different causal levels of is the only viable option that we have ready at hand (Boyd 1993, 2001). According to this view of reference and truth, both are epistemic achievements that depend upon the history of inquiry into the causal pathways that create local social and psychological phenomena, and where this inquiry rarely produces either perfectly exact knowledge or knowledge that can be usefully generalized to any novel circumstances whatsoever.

But let me conclude this chapter on a much less abstract point. In chapter 3, I introduced the distinction between enabling knowledge and psychological representations of structural content with a sports analogy, according to which enabling knowledge is the kind of expertise or know-how that—and this is the point—coaches know how to develop. Rarely do coaches instruct people on the rules. Instead, a good coach who is in the position of working with players who already know the rules of the game need only focus on discovering all the different ways that those players can become better and better at playing the game. I think the analogy between sports and social actions is more than just conceptually useful because it clarifies the notion of a regulative framework. What this analogy does, furthermore, is help make plausible the idea that normative moral psychology really could aim to provide something like evidence-based coaching for different forms of social life—if, that is, normative moral psychology has as its subject matter the forms of enabling knowledge that also count as functional virtues.

References

Alfano, Mark. 2015. "Can People Be Virtuous?—Ramsifying Virtue Theory." In *Current Controversies in Virtue Theory*, ed. Mark Alfano, 124–135. London: Routledge.

Annas, Julia. 2005. "Comments on John Doris's 'Lack of Character.'" *Philosophy and Phenomenological Research* 71 (3): 636–642.

Argyle, Michael. 1999. "Causes and Correlates of Happiness." In *Well-Being: The Foundations of Hedonic Psychology*, ed. D. E. Kahneman, 353–373. New York: Russell Sage Foundation.

Boyd, Richard. 1993. "Metaphor and Theory Change: What Is 'Metaphor' a Metaphor For?" In *Metaphor and Thought*, 2nd ed., ed. Andrew Ortony, 481–532. Cambridge: Cambridge University Press.

Boyd, Richard. 2001. "Truth Through Thick and Thin." In *What Is Truth?*, ed. Richard Schantz, 38–58. Berlin; New York: De Gruyter.

Burrow, Anthony L., Amanda C. O'Dell, and Patrick L. Hill. 2010. "Profiles of a Developmental Asset: Youth Purpose as a Context for Hope and Well-Being." *Journal of Youth and Adolescence* 39 (11): 1265–1273.

Burrow, Anthony L., Maclen Stanley, Rachel Sumner, and Patrick L. Hill. 2014. "Purpose in Life as a Resource for Increasing Comfort with Ethnic Diversity." *Personality and Social Psychology Bulletin* 40 (11): 1507–1516.

Diener, E. 2000. "Subjective Well-Being. The Science of Happiness and a Proposal for a National Index." *American Psychologist* 55 (1): 34–43. http://psycnet.apa.org/ (accessed April 2, 2015).

Diener, Ed, Eunkook M. Suh, Richard E. Lucas, and Heidi L. Smith. 1999. "Subjective Well-Being: Three Decades of Progress." *Psychological Bulletin* 125 (2): 276–302.

Doris, John M. 1998. "Persons, Situations, and Virtue Ethics." *Noûs* 32 (4): 504–530.

Doris, John, and Stephen Stich. 2006. "Moral Psychology: Empirical Approaches." *The Stanford Encyclopedia of Philosophy*, April 19. http://plato.stanford.edu/entries/moral-psych-emp/ (accessed November 22, 2014).

Dweck, Carol S. 2000. *Self-Theories: Their Role in Motivation, Personality, and Development.* 1st ed. Hove, UK: Psychology Press.

Elliot, A. J., and C. S. Dweck. 2007. *Handbook of Competence and Motivation.* 1st ed. New York: Guilford Press.

Erikson, E. H. 1994. *Insight and Responsibility*. New York: W. W. Norton.

Hill, Patrick L., and Anthony L. Burrow. 2012. "Viewing Purpose through an Eriksonian Lens." *Identity* 12 (1): 74–91.

Kahneman, Daniel, Edward Diener, and Norbert Schwarz. 1999. *Well-Being: Foundations of Hedonic Psychology*. New York: Russell Sage Foundation.

Lu, L. 1999. "Personal or Environmental Causes of Happiness: A Longitudinal Analysis." *The Journal of Social Psychology* 139 (1): 79–90.

Maria, W. Merritt, M. Doris John, and Gilbert Harman. 2010. "Character." In *The Moral Psychology Handbook*, ed. John Michael Doris, 355–401. Oxford: Oxford University Press.

Mischel, W., and Y. Shoda. 1995. "A Cognitive-Affective System Theory of Personality: Reconceptualizing Situations, Dispositions, Dynamics, and Invariance in Personality Structure." *Psychological Review* 102 (2): 246–268.

Mischel, W., and Y. Shoda. 1999. "Integrating Dispositions and Processing Dynamics within a Unified Theory of Personality: The Cognitive-Affective Personality System." In *Handbook of Personality: Theory and Research*, ed. L. A. Pervin and O. P. John, 208–241. New York: Guilford Press.

Park, Nansook, and Christopher Peterson. 2006. "Character Strengths and Happiness among Young Children: Content Analysis of Parental Descriptions." *Journal of Happiness Studies* 7 (3): 323–341.

Paunesku, David, Gregory M. Walton, Carissa Romero, Eric N. Smith, David S. Yeager, and Carol S. Dweck. 2015. "Mind-Set Interventions Are a Scalable Treatment for Academic Underachievement." *Psychological Science* 26 (6): 784–793.

Rödl, Sebastian. 2007. *Self-Consciousness*. Cambridge, MA: Harvard University Press.

Samman, Emma. 2007. "Psychological and Subjective Well-Being: A Proposal for Internationally Comparable Indicators." *Oxford Development Studies* 35 (4): 459–486.

Shimai, Satoshi, Keiko Otake, Nansook Park, Christopher Peterson, and Martin E. P. Seligman. 2006. "Convergence of Character Strengths in American and Japanese Young Adults." *Journal of Happiness Studies* 7 (3): 311–322.

Snow, Nancy E. 2010. *Virtue as Social Intelligence: An Empirically Grounded Theory*. 1st ed. London: Routledge.

Strauss, Anselm L. 1968. *A Time for Dying*. Piscataway, NJ: Transaction Publishers.

Verwijmeren, Thijs, Johan C. Karremans, Stefan F. Bernritter, Wolfgang Stroebe, and Daniël H. J. Wigboldus. 2013. "Warning: You Are Being Primed! The Effect of a Warning on the Impact of Subliminal Ads." *Journal of Experimental Social Psychology* 49 (6): 1124–1129.

11 Conclusion

We have now a theory of how moral psychology can be related to its auxiliary disciplines. The causal theory of ethics, described in chapter 7, provides a way of linking the conceptual resources of philosophical ethics, the empirical results of different social sciences and humanities, and new research questions in moral psychology. By applying the causal theory of ethics to the relevant social research, we can learn about which populations are appropriate for tests of local external validity in moral psychology, and thereby provide a way of validating research that is about any of the psychological questions that we encountered in chapters 4, 9, and 10. We have also seen how moral psychology can be divided into two separate areas of inquiry: the study of the psychological foundations of those social behaviors that are ethical, and the study of the characterological traits termed *functional virtues* in chapter 10, which are the forms of practical knowledge that allow a person to balance self-regard with prosociality. And the whole package stands in a form of methodological consilience with biology, insofar as it exemplifies a commitment to conceptual framework pluralism, and at the very heart of the package is a set of conceptual distinctions that are, technically, designed to ensure Mayr's lemma will apply to the study of human normative behavior.

This is a picture worth taking seriously because it is an example of how the social sciences, humanities, and moral psychology can cross-fertilize one another's research in a way that, just maybe, may lead to increased possibilities for metaphysical progress in ethics—but even for that, it is surely a superior picture to the alternatives that put philosophy, moral psychology, and the social sciences in competition over which single discipline gets to theorize morality.

Being "Biological" in Moral Psychology

The social turn in moral psychology is a turn away from a more popular approach to linking biology and moral psychology. The traditional answer says that "being biological" in moral psychology is to pursue the following strategy. First, describe a pattern of adaptive moral behavior in the EEA for either humans or some ancestral species. Then, formulate a hypothesis describing only one psychological mechanism that could have caused the relevant pattern of moral behavior. From this, *predict* that the psychological mechanism probably exists—and test this prediction by designing an experiment that looks only for evidence of the likely behavioral or cognitive effects of the posited mechanism. If these effects are observed, this is taken to confirm the prediction, and thereby validate the conclusion that the psychological mechanism in question probably exists. Moral psychology and biology are linked, according to this picture, only insofar as evolutionary theory can be a source of testable psychological hypotheses.

But we know that this is an unreliable pattern of inference. In evolutionary biology scientifically plausible proximate (that is: psychological, physiological, developmental, and so on) and ultimate (that is: structural) explanations stand in a many-to-many relationship. The mistake that fatally impairs the reliability of that pattern of inference is to assume only one hypothesis about a psychological mechanism that could have caused the relevant pattern of moral behavior. Mayr's lemma says that there will be multiple plausible psychological hypotheses that must be taken into account scientifically. So, determining which psychological mechanism could have caused a pattern of adaptive behavior in the EEA would require designing a study with conditions and controls that reflect the possible truth of *all* of the scientifically plausible proximate explanations that describe potential psychological causes of the relevant pattern of moral behavior. In most cases, such a study would be too complex to be feasible.

The alternative strategy for developing a biologically informed moral psychology is to analyze normative behavior in humans using an overall theoretical approach that is explicitly designed to be compatible with Mayr's lemma. That is achieved by adopting the technical distinctions between some regulative framework, the framework's structural content, psychological representations of its structural content, the framework's physical presuppositions, and any enabling knowledge associated with the framework's realization (chapter 3). We then assign to different behavioral (chapters 9 and 10) and social sciences (chapter 8) the task of investigating the phenomena that fall into these categories.

Universal Ethical Truths and Biology

We also need an ethical theory that avoids creating any inconsilience between philosophical ethics and moral psychology. The causal theory is just such an ethical theory. It defines a method of ethical inquiry that is not psychological: at bottom, the causal theory of ethics shows how investigating the population-level effects of both existing social institutions and interventions on these institutions can lead to the discovery of which norms are and are not ethical. But that is not the same as holding that ethical inquiry is evidentially neutral or statistically independent of psychological facts. Conclusions about the structural content of a regulative framework must be expressed in a different conceptual vocabulary than conclusions about how people understand its structural content. Ethical inquiry can proceed without using any nontrivial psychological premises, and yet it is not independent of human moral psychology. Ethics and moral psychology are relevant to one another in almost exactly the same methodological sense that, say, molecular biology and ethology are relevant to one another: both fields can inform one another, yet neither reduces to the other. Ethical inquiry and moral psychology should be pursued in parallel.

Unlike more traditional theories in philosophical ethics, the causal theory of ethics does not assume that an ethical theory must apply to all possible moral and ethical properties, or cover all possible normative phenomena. The theory is, by definition, restricted only to evaluating those norms that either are, or could be, embedded in existing social institutions. The causal theory of ethics therefore has nothing to say about, for instance, what a person should do if she is facing an entirely hypothetical choice between killing five people to save one thousand, or letting the thousand die to avoid choosing to kill five. But the theory is also inherently local: norms that are discovered to be ethical through application of the theory are not universal truths, nor are they truths that should compel the assent of any and all rational beings. In many cases, they will be rational only to the people whose behavior helps sustain the relevant social institutions.

For both of these properties, the causal theory achieves an even greater degree of consilience with evolutionary biology than it may at first seem. Evolutionary biology is a non-Newtonian science. It does not aim to discover absolute, necessary, universal biological laws of nature, and it is successful in part because of its ability to resist the search for such laws. Karl Popper put the relevant philosophical idea succinctly a half-century ago: "There exists no law of evolution, only the historical fact that plants and animals change, or, more precisely, that they have changed. The idea of a law that determines the direction and the character of evolution is a typical nineteenth century mistake, arising out of a general tendency to ascribe to the "Natural Law" the functions traditionally ascribed to God" (Popper 1959).

Popper's intent here is to point out that there are no laws in evolutionary biology that are like the kind of laws that God herself would presumably make: descriptions of universal regularities that cannot be accidentally broken or modified, and which, because they are known by an omniscient being, must also be necessarily true. And while in science there are many theories and models that employ sentences and equations that have the form of universal statements (All Fs are Gs), it is also now patently evident that these formulations are, at best, heuristic devices or approximations that are useful for describing local segments of some very complex natural process or system. They are not exact and literal truths (Giere 1999)—and they cannot be interpreted as if they are also objects of divine knowledge.

Yet contemporary philosophical ethics remains largely compatible with a Newtonian approach to normative truth. Moral philosophers can, without risk to the philosophical plausibility of their theories, assume that moral or ethical truths must be universal or necessary or categorical or intrinsically motivating or justified without recourse to empirical facts or what any rational person would agree with if they had all the evidence (for example, Shafer-Landau 1997; Hare 1981; Audi 2015)—which is to say that any moral philosopher who thinks about ethical norms as if they were functionally equivalent to divine commands does not put her professional standing at risk. So, why hasn't the failure of science to find evidence of any Newtonian laws of nature led to skepticism about the idea that ethical truths will turn out to be like Newtonian laws? Because a moral philosopher always can interpret the gap between the "oughts" of her ethical theory and the "is" of the world as a case of moral or ethical failure on the part of the world, moral philosophy can be very easily seen as in the business of producing regulative ideals that the world persistently undershoots. There is no reason to suspect that moral philosophers may be setting the standards defining knowledge of a "true" or "real" ethical norm too high.

In contrast, the causal theory of ethics is different and comparatively unique. Its method is a way of using patterns of reasoning that are characteristic of reliable scientific inquiry to confirm or disconfirm hypotheses about which norms are ethical norms, and where the method does not presuppose the existence of categorical ethical truths or other Newtonian laws in the normative sphere. At the limit, the causal theory of ethics may lead to the discovery that there are norms that, functionally speaking, are equivalent to such categorical or universal truths, because it is an empirical possibility that there may be a global normative context for which there is a set of norms that ought to be fundamental in that context. But the important point is that the concept *ethical norm* (as defined in chapter 7) does not requires there to be universal or necessary normative truths. Put differently, the definition of "ethical norm" at the center of the causal theory of ethics allows for the rational revision of which norms in which local contexts count as the ethical norms.

As I have noted, part of the cause of inconsilience between philosophical ethics and ADJDM moral psychology is the simple fact that the most plausible theories in philosophical ethics are nowhere realized (chapters 4 and 5). In practical terms, this means that moral philosophy has not provided moral psychology with a theory of ethics that is useful for determining which populations to sample from. The causal theory of ethics solves this problem: for example, by beginning from hard-question social theory, descriptive moral psychology can ask many different research questions about the psychological basis of the relevant behaviors. By distinguishing between descriptive moral psychology and normative moral psychology, we are then able to study the psychological links that bring together an individual and the social institutions constituting her larger social world (chapter 9), as well as the psychological basis (in both development and social cognition) of a person's relationship between her prosocial behaviors and the pursuit of her own personal values (chapter 10).

Reducing Epistemic and Moral Risk

If given a choice between a theory that looks for the normative equivalent of Newtonian laws of nature and a theory that defines ethical norms as rationally revisable truths that will rarely generalize without limit, there is in fact a positive argument for preferring the second option. Since scientific reasoning is, generally speaking, the best and most reliable form of reasoning that we have, then there is an argument for minimizing moral risk by minimizing epistemic risk that leads to some of the key ideas of the causal theory of ethics.

Here are the premises of the argument. The first is that the output of even the most rigorous and reliable scientific reasoning is, at best, only a collection of approximately true theories that are often found to be inconsistent with one another, and for which a great deal of expertise is required in order to know how to put any of these theories into action. The second premise is that, for two subarguments (one epistemic, the other moral), it is less risky to assume that there are moral facts than to assume that there are no moral facts. The first subargument goes like this. If there are no moral facts, then the assumption that there are moral facts will sooner or later be falsified by empirical research. But that cannot happen if it is both the case that there are no moral facts and we simply assume as much. In the latter case, instead of having empirical evidence of the absence of moral facts, we would only know that we simply assumed that there are no moral facts. So, whether or not there are moral facts, we end up with more knowledge if we adopt as a working assumption the view that there are moral facts, because the assumption will either be vindicated or falsified by scientific practice. The second subargument is this. It is morally less risky to assume the existence of moral facts than to assume that they do not exist. If moral facts exist and yet researchers in science accept as a working assumption that no such facts exist, then the institutions of science risk causing harm that would not otherwise occur were researchers to take on a different working assumption that there are moral facts. So, it is both morally and epistemically less risky to assume that there are moral facts. From that conclusion, we can infer that it is possible to minimize the moral risk of studying moral psychology scientifically by both assuming that there are moral facts, and just as importantly, by assuming that most of these facts can and should be studied by patterns of reasoning and inference that we have independent reason to think are as reliable as possible. These are of course the patterns of inference and explanation that are characteristic of the most rigorous forms of scientific inquiry—and, I stress, not just any form of nominally scientific inquiry whatsoever.

So, the conclusion is not that moral risk is minimized if we treat either ethics or morality as a science. Rather, the conclusion is that moral risk is minimized if we try to discover ways of using patterns of inference and explanation that are characteristic of reliable scientific inquiry to also carry out ethical and moral inquiry, and where we expect the outcome of applying these patterns to not exceed what is possible in the most mature scientific discipline. And one way of putting these recommendations into practice is social-turn moral psychology.

What about Choice, Freedom, Responsibility, Moral Motivation, and All the Psychological Phenomena That Are Traditionally within the Purview of Ethics?

We have noted several times now that understanding structural content requires only a conceptual vocabulary apt for describing structural processes in normatively integrated populations. This does not mean that there are not complex interactions between structural and psychological processes, and the details of descriptive and normative moral psychology amount to suggestions and guidelines about how best to study the psychological components of these interactions. But it does mean, as we have noted, that the causal theory of ethics is not a form of psychological inquiry.

This leads to an interesting objection: does that not show that the causal theory of ethics is not really an ethical theory? Or, if not that, that it is not a philosophically plausible ethical theory? The motivation for this objection is the idea that one of the most important ends of ethical inquiry—an end that helps to define standards that can be used to assess the plausibility of theories developed out of the practice of ethical inquiry—is to provide individual reasoners with choice and guidance about what they ought to do, practically speaking. But the causal theory of ethics cannot be used to do this. At best, it provides guidance to the various different practices and projects that have as their concern realizing increasingly widespread patterns of metaphysical progress in ethics. So it is a mistake to conclude that the causal theory of ethics is really an ethical theory.

The response to this objection involves another invocation of conceptual framework pluralism. It is true that the causal theory of ethics is not apt for providing individuals with guidance when they are making moral choices. But that does not mean that some other theoretical package cannot be used for that purpose. Indeed, I think it is plausible that some form of virtues ethics is most apt for understanding, and perhaps even guiding, what it is to make moral choices in everyday life. The deeper point here is to see that, for instance, virtue ethics does not compete with the causal theory of ethics. Indeed, we saw in chapter 10 how some of the key ideas in virtues ethics can be linked up with the causal theory of ethics—and so the overall package of ideas provides a way of combining concern with general interpersonal norms and their influence on social behavior with the world of personal values, character formation, and individual moral development. This combination is not only possible, but also desirable, because it is a sign of theoretical progress, from the perspective of conceptual framework pluralism. But all the same, it means that conclusions about ethical norms will not usually be derivable from conclusions about the formation of the capacities for love, sympathy, courage, and the other elements of a virtuous character.

More generally, then, the methodological recommendation is that moral philosophy would do well to resist the urge to use one single normative theory to address all important ethical and moral questions. Instead, moral philosophy can follow the lead of the biological sciences by seeking to assemble a self-consciously unsystematic and not logically organized collection of different theories that apply to different levels of social behavior—and consequently conceiving of epistemic progress as the ability to use different elements of this collection to yield increasing, deep understanding of human normativity. So, it is not that providing individuals with practice guidance is an unimportant end for normative theory—rather, the point is that we should not expect there to be only one philosophically adequate theory that is able to answer all important ethical questions, or otherwise satisfy all of the ends of ethical inquiry.

What about the Special Evidential Role of Intuitions in Moral Philosophy?

There is another important objection to social-turn moral psychology; here is the argument behind it. Moral philosophy, whether naturalist or otherwise, works the following way: philosophers take note of an initial stock of intuitions about moral issues, and they then make an attempt to systematize these intuitions. This is difficult work, which explains why, despite several generations of sustained effort, no ethical theory has emerged as the consensus view among moral philosophers. However, there is another way of understanding the source of the inconsilience between moral philosophy and ADJDM moral psychology. According to this alternative explanation, the patterns of inconsilience that are associated with, for example, definitions of *consequentialism* are really evidence that the correct definition of the concept is the one produced by ADJDM moral psychology. Why? ADJDM moral psychology, too, is an effort to systematize intuitions, but it uses the methods of cognitive and social psychology instead of the methods of philosophy. The theoretical picture which ADJDM moral psychologists have produced when they, too, have worked to systematize moral intuitions is, because of the superior reliability of scientific methods compared to the methods of philosophy, more likely to be the correct way of systematizing moral intuitions. If we must make a choice between intuition-based ethical theories developed by moral philosophers and the data-about-people's-intuitions-based ethical theories developed by ADJDM moral psychology, we should prefer the latter theories, again, because of the superior reliability of scientific inquiry.

This argument leads to the following objection: several of the key arguments in this book depend upon an ethical theory that is philosophical in nature. This theory is likely to be false, or at least less credible than the theories of ADJDM moral psychology, because ultimately it must somehow consist of an effort to systematize a certain number of intuitions—since that is just how ethical theories are justified, philosophically speaking. Consequently, inconsilience between moral philosophy and moral psychology shows that often philosophers get it wrong when they attempt to systematize intuitions, and since this book's ethical theory is also philosophical through and through, it too is likely mistaken.

But there is of course a good reply to this objection. In nearly all cases scientific methods are more reliable than the methods of philosophical inquiry, and therefore scientific efforts to systematize intuition are more likely to be reliable than philosophical efforts to do the same. But that does not matter, because it is a mistake to think that the only kind of evidence that matters for a philosophical theory is the theory's compatibility with a set of intuitions, however these intuitions are sorted, whose ever they are, and whatever their breadth may be. Readers are invited to note that the argument for the causal theory of ethics is almost entirely expressed in patterns of abductive inference, and where none of these abductive inferences attempt to summarize or systematize pretheoretical intuitions. No part of the argument for the causal theory of ethics—or indeed, any of the substantial claims made in earlier chapters—rests on justification that accrues because of evidence of a "fit" between the position in question and intuition. That of course explains why some moral philosophers may have a hard time seeing the argument for the theory; some may even complain that "it's not *really* philosophy" because the theory does not stand or fall purely as a function of its agreement with certain philosophical intuitions. (And we should always ask: *whose* moral intuitions exactly?) Nevertheless, since the inferences used to support the causal theory of ethics are mostly abductive inferences from empirical findings to theoretical claims, and since this pattern of inference is both ubiquitous in the natural and social sciences and often a characteristically reliable pattern of inference therein, it is a mistake to see the argument for the causal theory as a philosophical and *therefore not a scientific* argument. Ignoring the argument—and indeed, ignoring the existence of patterns of social behavior that are arguably ethical in nature—is tantamount to ignoring both scientific and philosophical evidence.

Final Words

Thus, social-turn moral psychology is not a turn away from either biology or philosophy; it is instead a way of reconfiguring the relationship between these two disciplines and empirical research in moral psychology that increases the latter's scientific rigor. The result is a complex—one might even say, untidy—picture. But that is intentional, because at one level social-turn moral psychology is an attempt to apply to the methodology of moral psychology Nancy Cartwright's answer to a question she posed nearly two decades ago: "How do we best put together different levels and different kinds of knowledge from different fields to solve real world problems, the bulk of which do not fall in any one domain of any one theory? [... How do we] develop methodologies, not for life in the laboratory where conditions can be set as one likes, but methodologies for life in the messy world that we inevitably inhabit?" (Cartwright 1999, 18).

Cartwright's answer is that the structure of scientific knowledge should messy because the world it represents is even messier still. Science should be seen as a loose confederacy of different but autonomous disciplines that develop their own proprietary conceptual vocabularies, patterns of cooperation and competition, standards of evidence and method, canonical theories, models, and formalisms, and so on—and this picture of science is the philosophical motivation behind social-turn moral psychology.

Bernard Williams is deservedly famous among philosophers for decrying the efforts of moral philosophy to understand morality by constructing a systematic theory of it. Among other costs, the move to systematize morality ensures that a moral philosopher spends her time thinking about something quite different from the real moral choices and actions of people, creating a situation in which moral theory simply floats above the complex and messy details of real life, and thereby never truly engages with it.

But there is one part of Williams's moral philosophy that has received much less attention, and that is the contrast he draws between ethics and science; despite the fact that, even though he was certainly quite right about the tendencies of moral philosophy, Williams got science rather badly wrong. He says "science has some chance of being more or less what it seems, a systematized account of how the world really is" (Williams 2011, 135). However, to a conceptual framework pluralist, nothing could be further from the truth. Efforts to systematize scientific knowledge will simply undercut themselves, because they will reduce the number of natural phenomena that we can understand in causal terms. In order to preserve their (often very hard won) ability to make reliable counterfactual-sustaining inferences about complex natural system, scientists really must be comfortable making continual trade-offs between descriptive breadth, accuracy, explanatory depth, mathematical tractability, pragmatic utility, and so on—it is only metaphysicians who attempt systematicity; "scientists pursue more local goals, at several layers of increasing locality" (Wimsatt 2006). The contrast that Williams draws between science and ethical inquiry misses the mark, thus. Science is not in the business of producing a systematic account of how the world really is, and it would cease to be successful if it tried.

Still, there is a deeper point in the neighborhood here. Social-turn moral psychology is a way of asking and then trying to answer some of the traditional questions of moral philosophy without also taking on a number of the methodological and conceptual elements that Williams critiques. For example, Williams writes:

> [Ethical] theory looks characteristically for considerations that are very general and have as little distinctive content as possible, because it is trying to systematize. ... The only serious enterprise is living, and we have to live after the reflection; moreover (though the distinction between theory and practice encourages us to forget it), we have to live during it as well. Theory typically uses the assumption that we probably have too many ethical ideas, some of which may well turn out to be mere prejudices. Our problem now is that we have not too many but too few, and we need to cherish as many as we can. (Williams 2011)

But the package of ideas that constitute social-turn moral psychology deliberately imports some of the untidiness of human normative life into our theory of it, and it resists the urge to find a further and more general level of abstraction that can contain the untidiness by constructing increasing abstract, increasingly general theories. Someone committed to the conceptual framework pluralism built into social-turn moral psychology can happily make room for as many ethical ideas and as much distinctive content as can be found—because they must also be satisfied with an intellectual framework that does not aim to provide either the ethical expert or the moral person with a complete, total, and unified picture of how to go on, whether in theory or in real life.

We have gotten past the idea that science will tell us once and for all how nature really is; perhaps it is time to ask what it means to hold ethical inquiry to exactly the same standard.

Reference

Audi, Robert. 2015. "Intuition and Its Place in Ethics." *Journal of the American Philosophical Association* 1 (1): 57–77.

Cartwright, Nancy. 1999. *The Dappled World: A Study of the Boundaries of Science*. Cambridge: Cambridge University Press.

Giere, Ronald N. 1999. *Science without Laws*. Chicago: University of Chicago Press.

Hare, R. M. 1981. *Moral Thinking: Its Method, Levels, and Point*. Oxford, UK: Oxford University Press.

Popper, Karl R. 1959. "Prediction and Prophecy in the Social Sciences." In *Theories of History*, ed. Patrick Gardiner, 276–285. New York: Free Press.

Shafer-Landau, Russ. 1997. "Moral Rules." *Ethics* 107 (4): 584–611.

Williams, Bernard. 2011. *Ethics and the Limits of Philosophy*. London: Routledge Classics.

Wimsatt, William C. 2006. "Reductionism and Its Heuristics: Making Methodological Reductionism Honest." *Synthese* 151 (3): 445–475.

Index

Adaptation, 23–24, 27–28, 33–34, 42, 76
Adaptive behavior, 27, 34, 128, 220
Adaptive rationale, 27, 46
ADJDM experiments, 93
ADJDM moral psychology, 92–94, 96–97, 105, 110–112
ADJDM research, 89, 91–95, 97, 99, 101, 103, 105, 107, 111, 117–118, 133, 182, 184–186, 197–198, 227
Adulthood, 187–188
Aerial predators, 29–30, 48
Affective responses, 108, 189–190
Age, 119, 138, 158, 165, 206
Agency, 7, 211
Aggregations, 148
Analysis
  conceptual, 49
  levels of, 5, 29, 38, 143
  population-level, 181
  statistical, 197
  structural level of, 5, 181
Analytical concepts, 16, 49, 55–56, 188
APA's *Ethical Principles in the Conduct of Research with Human Participants*, 164–165
Approximations, 7, 13–14, 17, 73, 83, 89, 178, 222
A priori inferences, 181
A priori truths, 45, 84
Argument, 3–5, 7–11, 24–25, 44, 71, 74–75, 79–80, 117–18, 128, 143–144, 149, 153–154, 210–212, 224, 226–228
Aristotle's ethics, 211
Association for Psychological Science, 47, 114, 130
Assumptions
  background, 94, 134, 173
  methodological, 69, 173
Attitudes, 19, 29, 51, 54, 62, 72, 77, 102, 194
Autonomy, 58, 187–188, 200–201

Behavior, 16–17, 23–29, 31–34, 36–37, 39–41, 43–45, 104, 106–107, 171–172, 174–175, 184–186, 192–193, 195–198, 206, 209–210
  altruistic, 194
  cooperative, 43
  coordinated, 12
  helpful, 41
  norm-regulated, 50, 206
  prosocial, 40, 223
  proxy, 197
  real-world, 184
  relevant, 34, 49, 53, 176, 185, 198, 223
Behavioral ecology, 28, 48
Behavior generators, 26–28
Being biological, 24–25, 220

Beliefs, 3, 19, 38, 51, 54, 62–65, 72, 75–76, 81–82, 95, 102, 111, 174, 176
ethical, 82
individual, 152
normative, 39
prior, 11
Biological sciences, 28, 45–47, 69, 226
Biology, 13, 23–27, 29, 31, 33, 35, 37, 39, 41–43, 45–47, 86, 130, 219–221, 228
contemporary evolutionary, 35, 128
molecular, 221
Biopower, 161–162, 179
BRAC, 165–167, 179

Canada, 114, 184
Categorical variable, 126
Categories, 19, 23, 28–29, 43, 76, 81, 124, 126–127, 154, 214, 221
exclusive, 81
higher-level, 127
intermediary, 188
psychological, 97
Category mistake, 77, 97–98, 100, 103
Catholic norms, 142
Causal diagrams, 157–158
Causal graphs, 186
Causal influence, 84, 86
Causal interactions, 84, 86, 139, 160
complex, 172
Causal processes, 135, 161–162, 185
Causal properties, 170–171
Causal structures, 119, 170
Causal systems, 11, 170–171
Causal theory of ethics, 10, 133, 135, 151, 169, 174, 192, 221–222, 228
Causal unity, 163, 192–193
Chain, inferential, 90
Characterological traits, 207–208, 211, 215, 219, 226
Character traits, 39, 206, 208, 212
Chemistry, 27, 86
Childhood, 187–188
Child mortality, 138
Children, 42, 52, 56, 138, 161, 165, 167, 178–179, 194–195, 200, 206–207
Choices, 12, 16, 36, 49, 58–59, 70, 94–95, 111, 122–123, 126–127, 142, 147, 183–184, 224–225, 227
hypothetical, 221
Clusters, 9–10, 52–53, 141, 144, 149, 155, 175, 188–189, 192, 195
Coaches, 56, 216
Code, tax, 71
Coding, scientific, 120, 124
Cognition, 20–21, 86, 112–13, 133, 182, 192
Cognitive science, 47, 83–84, 115, 131, 201
Cognitivism, 80–82, 107, 109–110, 117
Community, 4–6, 17–18, 52, 63, 96, 106, 108, 148, 150, 187–188, 193, 200–201
normatively integrated, 4, 11, 17, 190
Complexity, 13, 31, 69–70, 144, 171–173
Complex social behaviors, 50–51, 197
Components
conceptual, 64, 135
psychological, 205, 225
Conceptions, 39, 60, 142, 144–148, 176, 182, 197, 213
Concepts, 8–10, 16–18, 27, 51, 54–55, 63–65, 69–70, 89, 120, 141–143, 161–163, 169–170, 208–209, 211–212, 214–215
deontic, 100
entrenched, 133
normative, 90–91, 104, 106, 215
Conceptual distinction, 106, 144
Conceptual framework pluralism, 10, 13–16, 18–19, 26, 35, 45, 91, 153, 155, 196, 213, 219, 226, 229–230
perspective of, 19, 206, 226
Conceptual frameworks, 14, 16, 61, 64, 66, 161, 176, 211

Conceptual practices, 117, 142
Conceptual resources, 135, 219
Conceptual vocabularies, 4–5, 10, 13–16, 28, 55, 65, 69, 91, 143–144, 164, 170, 215, 221, 225, 229
alternative, 134
single, 15, 45, 144, 196
technical, 196
unified, 45
Confirmation, 32, 47
scientific, 28, 30, 74, 149
Conflict, 4, 11–12, 36, 99, 106, 126, 142, 175
Conscience, 94–95
Consequentialism, 66, 80, 97–99, 102–103, 115, 211, 227
Consilience, 8–9, 44, 70, 117, 222
Constructivism, 81, 83, 85–86, 107, 109–110
Constructs, experimental, 100
Container test, 185–187, 189–192, 195, 197–199
Cooperation, 12, 36, 40, 43–44, 48, 107–108, 178, 205, 229
large-scale intergenerational, 206
social, 4
Cornell University, 2
Counterfactual dependence, 54
Creativity, 93, 173
Criteria, switching, 151
Critique, 20, 43, 46, 67–68, 128, 163

Darwin, Charles, 40, 170
Debunking, 199
Decision procedure, 58–59, 74, 99
Decisions, 59, 92–93, 99, 101, 103, 109, 143, 147, 158, 183, 185, 198
Decision strategies, 104
Decriminalization, 169
Deliberation, 16, 94, 108, 118, 183
interpersonal, 94
mediated, 111
social, 92
Democracies, 45
Deontology, 15, 55, 63, 66, 75–77, 79–80, 97–104, 113–114, 124
Descriptive moral psychology, 10, 159, 181–182, 185–186, 191–192, 195–196, 198, 206, 223
Developmental interventions, 150
Dignity, 127, 142, 146, 174
Dilemmas, 6, 91–93, 99, 101, 106, 111, 121, 123, 126
methodological, 133
trolley, 94, 97
Disciplines, 1, 6, 8, 65, 111, 153, 219, 228
autonomous, 229
auxiliary, 1, 219
Discoveries, 47, 169, 171, 221, 223
moral, 110
Divinity, 189–190, 200–201
Domains, 27, 37, 110, 119, 121, 129, 173–174, 188, 228
ethical, 110, 206
moral, 159, 169, 188–189, 205
Dumbfounding studies, 109–110, 119, 121, 125, 127–128
Duties, 39, 57–59, 62–63, 84, 127
Dynamics, 9, 38–39, 201, 217
structural, 36

Economic analysis, 173, 177–179
Economic growth, 19, 146, 158
Economics
behavioral, 160
positive, 175
Economic structure, 106–107
Economists, 19, 172–173, 175
Education, 165–166, 169, 178–179, 196, 209
Education programs, 147, 165–167, 192–193, 209

EEA, 25, 34, 43, 198, 220
Effects
actual, 152
behavioral, 33, 175
causal, 81, 83, 151–153, 163, 175
good, 11–12, 152
local, 214
multiple, 204
observed, 150
population-level, 198, 221
structural, 152
unintended, 204
Either/or choice, 92, 94
Elements
conceptual, 111, 230
constructivist, 86
intuitive, 111
Emotions, 82–83, 114–115, 130
Empirical research, 2, 6, 42–43, 134, 136, 155, 157–158, 177, 206, 224, 228
Enabling knowledge, 55–56, 58–59, 64, 73, 134, 143, 181, 203–205, 214–216, 220
Environment, 23, 25, 31, 34, 147, 183, 198
Epistemic goals, 14
Epistemic progress, 170, 226
Epistemic virtues, 168
Epistemology, feminist, 108
Ethical behaviors, 2, 6, 10, 44, 152, 156, 169, 172, 175–176, 186, 192, 195–196, 204, 206, 210
actual, 6
human, 63
observed, 185
patterns of, 176, 181, 186, 191, 195, 204, 207
real-life, 182, 185
theory of, 96
Ethical beliefs and judgments, 82
Ethical ideas, 230
Ethical inquiry, 107–108, 136, 151–152, 160, 177, 198, 221, 225–226, 230
rational, 109, 152, 155
Ethical knowledge, 16, 103, 118, 168, 197
picture of, 196–197
Ethical norms, 12, 59, 63–64, 66, 141–142, 151–155, 169, 172, 214, 222–224, 226
cluster of, 141, 167
formal, 165
general, 152
necessary, 154
well-justified, 63
Ethical progress, 15, 108, 168
Ethical properties, 81–83, 86, 221
Ethical regulative framework, 66, 69–70, 74, 99, 104, 134, 141
Ethical theories, 6, 9, 59, 62–63, 67, 70–72, 76–79, 97, 101–104, 124, 154–155, 172, 221–222, 225, 227
families of, 72, 75, 79
ideal, 61
intuition-based, 227
novel, 103, 105
philosophically plausible, 78, 98, 225
plausible, 78, 98, 105
Ethical truths, 222
categorical, 223
Ethics, 49, 57–59, 61–63, 67–71, 83–85, 89–91, 105–107, 111–113, 133, 153–155, 157–159, 168, 199–200, 225, 229–231
causal theory of, 10, 18, 133–135, 142–144, 152–153, 155, 160, 168–173, 176–177, 196–197, 213–214, 219, 221, 223–226, 228
human, 69
religious, 66, 70
theory of, 6, 10, 15, 70, 223
Ethics and moral psychology, 90, 221

Ethnography, 153
Ethology, 46, 48, 221
Euthanasia, 183–184
Evidence, 5–6, 11–14, 35–36, 39–40, 42–44, 86–87, 89–91, 104–105, 109–110, 127–128, 135–136, 155–156, 186–189, 192–193, 227–229
  psychological, 117
Evolution, 20, 23, 26, 35, 37, 41, 43, 47–48, 112, 129, 199, 222
Evolutionary anthropology, 140
Evolutionary biology, 1, 8–9, 23–26, 28, 35–36, 42, 44, 96, 128, 134, 143–144, 157, 177, 220, 222
Evolutionary explanations, 8, 24–26, 39
Evolutionary game theory, 37
Evolutionary theory, 1, 25, 29, 96, 129, 198, 220
Experimental protocols, 93–94, 126
Experiments
  breaching, 16–17
  natural, 169
  scientific, 16, 79–80, 83–85, 92, 94, 105, 118–121, 125, 163, 165, 182, 194, 208, 212, 220
Expertise, 14, 16, 110, 155, 160, 216, 224
  actual existing, 108
  certified, 108
Experts, 16, 83, 92, 94, 146
  ethical, 160, 230
  general, 160
Explanations, 13–14, 16, 19–20, 23–29, 31–34, 36, 40, 44, 89–90, 92, 107–108, 111, 138, 155, 227
  alternative, 122, 204, 227
  confirmed, 34
  extrapolative, 128
  individualistic, 18–20, 24
  levels of, 4, 26
  patterns of, 25, 170
  plausible, 34, 44, 122, 138
  verbal, 119
External validity, 6, 18, 89–91, 93, 111–112, 128, 133, 155, 159
  local, 93–94, 96, 134, 156, 169, 185, 187–186, 190–191, 199, 219
  universal, 93, 96, 104, 134, 182, 196

Failure, ethical, 223
Fairness, 81, 168
  justice as, 67
Families of ethical theory, 72, 75, 77, 97, 99, 101–102
Feedback loops, 145, 166, 199
Fitness, 23, 33, 42
Fixed mindset, 208
Formal theories, 171, 173–174, 229
Framework, 8, 16, 51, 53–57, 63–65, 70, 91, 112, 155, 175, 17, 2220
Freedom, 146, 225
Functional virtues, 203–216, 219
Functions, 11–13, 17, 21, 25, 27, 60, 92, 108, 136, 223, 228
  adaptive, 38
  evolutionary, 26

Games, 9, 17, 49–56, 65, 76, 201–203, 214
GapMinder, 136, 138, 157, 181
Gaze, 182–184
GDP, 137, 147
General principles, 102, 197
Genetic drift, 23
Genetic programming, 38
God, 224
Good and valuable outcomes
  new, 168
  particular, 139
  society-wide, 193
Good and value outcomes, 139, 144–147, 151, 154–155, 166–169, 176, 181, 184–185, 188, 191–193, 196
Good outcomes
  small-scale, 195
  small-scale social, 192

Goods
basic, 138, 141
personal, 193
Gould, S. J., 42
Groups, 4–6, 40, 56, 59, 63–65, 73, 76, 80–82, 98, 125, 127, 140, 161–163, 188–191, 194–195
Growth, equitable, 19
Growth mindset, 208–209

Happiness, 57, 63, 161, 178, 205, 207–208, 216–217
Hard questions, 10, 152, 156, 159–161, 163, 165, 167–169, 171, 173, 175–177, 180, 185–186, 188, 191–192, 205
Hare, 142, 157, 222, 230
Heinz dilemma, 109, 122, 126
Heuristics, 36, 51, 231
general, 36
History, 4–5, 11, 28, 35, 45, 47, 52, 76, 100, 112, 143, 155, 172, 215, 231
ethical, 5
evolutionary, 1, 29, 35
intellectual, 107
Hockey, 50–51, 54–56, 76, 104, 203, 205
regulative framework of, 54, 56, 76, 141
Hockey players, 54–55, 76
Human behavior, 24–26, 46, 114, 153–155, 181, 200
patterns of, 24–25, 32, 172, 193
Humanities, 156, 162, 219
Human mind, 86, 94, 129, 171, 213
Human nature, 58, 197, 199
Human normative behavior, 15–16, 63, 91, 108, 159, 213, 219
Human populations, 5, 94, 96, 135–136, 214
existing, 6
integrated, 145
Hypotheses, 27, 32, 38, 41, 83, 85, 93–94, 100, 104, 107, 110, 176, 184, 186, 220
auxiliary, 34
behavioral, 28
comparative, 110, 121
conceptual, 136
disconfirm, 223
favored, 85
metaphysical, 86
neurological, 107
normative, 119
physiological, 55
scientific, 25, 41, 134–135
testable, 197

Ideologies, 72, 147, 153, 161–162
Ignorance, 60, 62, 74, 164
Impact evaluations, 150, 158
Inconsilience, 8–10, 35, 89–91, 93, 95–97, 99, 101, 103, 105, 107, 109, 111, 117, 155–156, 227
Individualism, 19–20
Individuals, 19, 29, 72, 107, 118, 148–149, 151, 153, 183, 189, 205, 226
Induction, 45, 69, 87, 90, 154
Inferences, 32, 37, 39–40, 83, 90, 96, 103–104, 111, 125, 128, 147, 171, 220, 224–225, 228
abductive, 139, 228
causal, 150, 157
extrapolative, 96, 99
predictive, 34
Information, 1, 13, 20, 150, 171, 175, 177, 183, 189, 192, 194
causal, 175
genetic, 27
normative, 2
Innate ideas, 84
Innocent people, 101
Inquiry, 7, 10, 14, 45, 50, 66, 151, 154, 176–177, 215, 219, 225
pluralistic, 129

Insufficient justification, 124
Intelligence, 210, 212
social, 212–213, 218
Interactions, complex, 50, 143, 225
Interdependence, 146, 208
Interdisciplinary, 162, 196
Interests, 26, 38, 60–61, 125, 128, 138, 164
Interventions, 150–152, 156, 171, 209, 221
causal, 150
Intuitionism, 81–86, 105, 110
ethical, 38, 62, 87
Intuitions, 6, 38, 59, 62–63, 83–84, 86–87, 120, 123, 128–130, 187, 190–191, 226–228, 230

Judgments, 2, 81–84, 91–93, 97–98, 100, 102–104, 108–110, 118, 122–123, 125, 190, 194
adult deontic, 92
consequentialist, 98
hypothetical, 98, 199
intuitive, 106
moral domain, 119
non-moral domain, 121
non-utilitarian, 101
normative, 98
patterns of, 93, 99
psychological explanations of patterns of, 93, 99
sampled, 91
unchanging, 108
utilitarian, 100
Judgments of philosophical plausibility, 84–85
Justice, 61, 67–68, 81–82, 146
principles of, 61–62
Justifications, 12, 47, 50, 75, 98, 104, 114, 119, 124–125, 130, 154, 228
articulate, 108
credible, 123
imperfect, 125
inadequate, 124
Just-so stories, 42–43

Kant, Immanuel, 49, 58–60, 62, 70, 74–75, 77–79, 114, 186
Kant's ethics, 58, 67, 75–77, 79, 104, 128, 211
Knowledge, 13, 54, 56–57, 60–64, 84, 86, 94, 97, 150–151, 155, 169–170, 197, 203–206, 212, 215
applied, 83
divine, 222
local, 171
scientific, 166, 176, 229
unconscious, 123

Laws, 19, 50, 58–59, 68, 121, 126, 154–155, 167–168, 174, 222, 230
universal, 58, 170, 172
Laws of human behavior, 154–155
Levels, 3–5, 45, 110–111, 136, 138, 140, 157–158, 166–167, 174–175, 181, 188, 192, 226, 228, 230
causal, 215
lower, 33
multiple, 29
Lewontin, R. C., 42
Liberals, 94, 101, 130
Life expectancy, 136–137, 139
Linguistic analogy, 130
Load, cognitive, 109
Logic, 38, 71, 75
Longevity, 139
Longitudinal analysis, 217

Many-to-many relationship, 28, 32, 177
Mass media, 188–189
Mayr's lemma, 8–9, 31–35, 38, 42, 44–45, 50, 96, 134, 143, 219–220
violations of, 32–33, 39
Measurement, 146–147, 158

Mechanisms, 26–27, 29, 31, 33, 50, 120, 186, 196–195, 197
mental, 38, 40
Metaethical neutrality, 109–110
principle of, 105, 108–109
Metaethical positions, standard, 80, 85
Metaethical theories, philosophically plausible, 85, 97
Metaethics, 79–81, 85, 129
contemporary, 80, 83, 85, 97
Metaphysical change, 169
Metaphysics, 67, 84, 90, 144–145, 149
Methodological consilience, 9, 143–144, 219
Methodological individualism, 19, 63, 66, 96, 159
Methodological norms, 69, 71, 73, 75, 77–79, 81, 83, 85, 87, 92–94, 97, 151, 153–154, 165, 228–230
Microfoundations, 10
Microsystem, 182, 191–192, 195, 197–198
Mill, J. S., 1, 49, 57–58, 62, 71, 74–75, 77–79, 98, 125, 131, 186
Model fitting, 175
Models, 13–14, 20, 36–37, 48, 69, 108, 170–175, 178, 183, 186, 188–191, 196, 199, 222, 229
approximate, 171
formal, 171
mathematical, 172
multilevel, 153
mutually inconsistent, 14, 170
statistical, 93, 157
synthetic control, 167
Modules, 25, 108, 190
evolved moral, 39, 108, 190
Moral anti-realism, 80
Moral behavior, 25, 35–37, 43–44, 96, 114, 128, 169, 197, 220
adaptive, 220
contemporary, 43
Moral choices, 185, 226
Moral cognition, 82–83, 86, 91, 93, 96, 102, 105, 108, 111, 119–120, 124–125, 127–128, 131, 186–187, 189
Moral decisions, 183–185
Moral development, 127, 211, 226
Moral dilemmas, 99, 112–113, 121, 123, 186
Moral dumbfounding, 110, 117–123, 125, 127–131, 198
Moral facts, 108, 224
Moral instincts, 39, 47, 128
innate, 127–128
Moral intuitions, 39, 62–63, 70, 92, 102, 106–108, 111, 127–129, 190, 227–228
instinctive, 40
Morality, 10–13, 15, 21, 25, 35–37, 39–43, 47, 67, 69–70, 96, 99, 113–114, 157, 200–201, 225
evolution of, 35, 39–41, 47, 96, 198
first principle of, 77
human, 37
mysteries of, 98, 113
rationality of, 11, 13
sense of, 39–40
systematize, 229
Moral judgments, 2, 21, 38–39, 41–42, 47, 80, 82, 87, 93–94, 100, 102, 113–114, 117–119, 121–128, 130–131
conceptualized, 39
etiology of, 122, 124
generating, 125
individual, 38
intuitive, 123
ordinary, 118
Moral knowledge, 11, 13, 58, 186
Moral norms, 36, 143
traditional, 15
Moral philosophers, 9, 17, 39, 71, 75, 77, 106, 129, 142–143, 152, 222–223, 227–229
high-status, 108

Moral philosophy, 8, 15, 59, 77–78, 100–101, 118, 128–129, 133, 223, 226–227, 229–230
contemporary, 15, 18, 106, 197
history of, 6, 142
Moral principles, 102–103, 115, 118, 123
Moral problems, 94, 119, 121, 130
intractable, 5
real, 94
Moral properties, 117–118
Moral psychologists, 5, 9–10, 18, 32, 44, 62, 69, 75, 78, 85, 90, 159, 169, 172, 175–176
Moral psychology, 1–2, 4–10, 66–67, 76–79, 90–94, 96–97, 105, 110–115, 117–118, 127–129, 133–135, 184–187, 195–201, 219–221, 227–228
microsystem, 192
Moral rationalism, 87, 102, 118, 127
Moral reasoning, 108–110, 118
Moral relativism, 174
Moral risk, 224–225
Moral sense, innate, 128
Moral Sense Test, 123–124
Moral sense theories, 39, 41–42
Morning sickness, 33, 46
Motivation, 33, 36, 53, 58, 62, 65, 82–83, 97, 102, 145–146, 208, 210, 212, 217, 225
Multilevel theory, 186

Naturalism, 80–81, 83, 85–86, 107, 109–110, 117
ethical, 10–13, 16, 86
Natural selection, 8, 23–24, 27–28, 31, 33–34, 38–43, 69, 76, 143, 158
Nepotism, 29, 48
New synthesis, 37, 39, 47–48, 114, 200
Newtonian laws, 70, 222–224
Nominalism, tactical, 135–136, 149, 192, 214
Noncognitivism, 80, 82–83, 85, 105, 110
Non-ideal theory, 61
Norm, single, 12, 53, 73, 75, 142
Normative behavior, 9, 14–16, 69, 91, 155, 196, 220
dimension of human, 63, 159
Normative contexts, 141–142, 151–152, 166–167, 176, 203–205
global, 223
independent, 150
Normative ethics, 78, 80, 104, 199
Normative moral psychology, 181, 199, 203, 205–217, 223
Normative structure, 7, 63, 135
Normative theories, 2, 70, 73–77, 80, 97, 99, 102, 131, 171, 173, 197, 214, 226
correct, 15
guiding, 77
ideal, 61
Normative truths, 16, 222
Norms, 11–12, 16–18, 50, 52–57, 59–60, 63–66, 69–71, 139–143, 151–152, 154–155, 166–167, 175–176, 198, 221, 223
causal effects of their realization, 166, 169
clusters of, 52, 56, 64, 139, 141, 149, 153, 155, 165, 168–169
embed, 190
epistemic, 16
ethical, 63
explicit, 49
fundamental, 4, 59–60, 70, 110, 214
general, 168, 226
ideal, 17–18
individual, 53
individuating, 175
interrelated, 53
legal, 3, 167
local, 94, 169–170
nonethical, 142

Norms (cont.)
obligatory, 13, 206
official, 54
prescriptive, 165
realized, 16–18, 140
social, 36, 67, 140, 161, 194
universal, 166
unrealizable, 17–18
unrealized, 103

Obedience, 127, 163, 177–179
Objects, 10, 81, 84, 172, 222
Obligatory, 12, 65, 194
Occupation, 2–4
OECD, 147, 158
Oppression, 161–162
Outcomes, 4–5, 7, 99–100, 103, 135–136, 144–145, 147–148, 150–151, 154, 161–162, 166, 181–183, 188, 190, 3
bad, 168–169, 212
balancing existing valuable, 168
better, 152, 156, 159, 165
detrimental, 159
ethical, 167, 190, 209, 214
intervention, 150
particular, 135
patterns of, 148, 154, 172, 204
positive, 208
ranking, 99
realized good and valuable, 195
structural, 6
worse, 140

Pairs, ultimate-proximate explanations, 40
Path-dependency, 18
Patterns, 9, 11–12, 17–18, 24–25, 31–32, 43–44, 90–91, 122, 155–156, 161–162, 182–185, 195–196, 204–205, 220, 227–229
adaptive, 26, 128
behavioral, 122
ethical, 2
particular, 34, 140
real, 98, 108, 172, 182
response, 122
Payoffs, 99
Personal character, 150
Personal outcomes, 181, 208
Philosophers, professional, 70, 80, 106, 108, 160
Philosophical ethics, 1–2, 6, 8–10, 49, 63–66, 70–72, 77–78, 89–91, 96–99, 102–107, 111–112, 117–119, 124–125, 128, 156
contemporary, 18, 106, 128, 222
core concepts of, 67, 90
history of, 57, 64, 69, 79
perspective of, 90, 104, 128
plausible theories in, 10, 49, 74, 77–78, 97, 128, 223
Philosophical ethics and moral psychology, relation between, 8–9, 49, 76, 90–91, 111, 117, 133, 221
Philosophical evidence, 156, 228
Philosophical plausibility, 6, 79–80, 84–85, 112, 134, 222
Philosophy, professional, 106–107, 160
Physical presuppositions, 53–58, 61–64, 70, 72–77, 134, 181, 220
Policies, educational, 166–167
Political economy, 1, 20, 158, 177–178
Population-level causal process, 158
Populations, 4–7, 10, 18–19, 23, 64, 66, 94, 96, 136–139, 143, 145–147, 167, 169, 181, 186–187
national, 139
normatively integrated, 4–7, 74, 96, 136, 139, 144, 155, 159–160, 168–169, 175, 181, 188, 192–193, 225
transient, 187
Poverty, 150, 157, 168
Preferences, 101–102, 145, 148, 173, 175–176
individual, 194, 194
ordered, 104

Presuppositions, 54, 72, 74–75, 77–78, 97, 143, 191
Principles, 8–9, 59–64, 70–71, 74, 77–79, 85, 89–90, 100–105, 115, 118, 123, 129, 133, 154
ethical, 103–104
normative, 129, 194
Procedural fairness, 5, 127
Processes, complex population-level, 136
Progress, metaphysical, 170, 201, 219, 225
Projectability, 84, 90, 149
Properties
basic, 176
characterological, 63
computational, 183
constructed, 81
factual, 124
logical, 28
material, 81
political, 103
referential, 169
stable, 188
structural, 31
supra-individual, 18
technical, 89
universal, 94
Property rights, 70, 126
Propositional content, 2, 17, 54, 57, 82, 144, 152, 175
Prosocial, 204–206
Prostitution, 167
Proximate, 25–32, 40–41, 50, 64, 143–144, 195
Proximate and ultimate explanations, 26, 30–32, 40
Proximate explanations, 23–24, 27–28, 30–32, 34, 37–41, 43–45, 55, 176
confirmed, 32
independent, 39
plausible, 28, 30, 32, 41, 44, 220
scientifically plausible, 30–32, 34, 44, 220
true, 27
Proximate hypotheses, 27, 32, 42
plausible, 38
single, 38, 40
Proximate mechanisms, 26, 31, 33–34, 37–38, 40–41
cognitive, 29
complex, 40
discrete, 34
endogenous, 43
particular, 33
unknown, 34
Proximate phenomena, 30, 32
Proximate processes, 31–32, 49–50
Proximate questions, 30
Proximate-ultimate distinction, 28–29, 31–32, 35, 46
Psychological basis, 5, 74, 190, 206, 223
Psychological conclusions, 8, 24–26, 36, 119
Psychological content of a regulative framework, 51, 55, 64, 72–74, 143, 189, 191
Psychological experiment, 8, 71–72, 75, 86, 165, 209
Psychological explanations, 5, 8–9, 19, 23, 44–45, 93, 99, 184
Psychological foundations, 7, 9, 77–79, 97, 149, 182, 192, 206, 214, 219
Psychological hypotheses, 1–2, 55, 65, 85, 176, 185, 194, 210
Psychological mechanisms, 24–25, 27, 39, 43, 89, 93–94, 184, 220
Psychological model, 186–187
Psychological premises, 71, 75, 143
Psychological processes, 2, 107, 120, 186, 189, 225
dissociable, 106
Psychological representations, 51, 54–59, 62–66, 69, 72–77, 97–98, 102, 134, 176–177, 181, 203, 220

Psychological representations of structural content, 176, 216
Psychological research, 65, 78, 117, 181, 192–193, 199
Psychological theories, 70, 89, 103, 134, 182, 184–185, 187, 191
Psychologists, social, 39, 211–212
Psychology, 1–2, 23, 47, 65, 72, 86–87, 89, 91, 95–101, 113, 181, 191, 197, 201, 210–211
  cognitive, 55, 195
  developmental, 32, 42, 113, 195, 200
  evolutionary, 32, 38, 47, 128
  human, 26, 75, 96, 200, 214–215
Punishment, 38–39, 126, 210

Rape offenses, 167
Rational agents, 58, 83
Rawls, John, 59–62, 64, 71, 74–75
Realism, moral, 80
Realizable norms, 16–18, 50
Realization, 54, 66–67, 69, 72–74, 98, 103–104, 129, 135–136, 139–141, 144–145, 147, 149, 186, 193–194, 203–205
Reasoning, 40, 53, 75, 98, 103–104, 107–109, 114, 117–119, 122, 125, 127, 142, 190, 198, 224
  deontological, 124
  patterns of, 103, 223–224
Regularities
  biological, 170
  causal, 174
  general-law-like, 154
  local behavioral, 197
  universal, 222
Regulative frameworks, 51–61, 63–67, 70–76, 97–99, 134, 139–141, 143, 149, 152, 154, 166, 168, 175–177, 203–204, 220–221
  existing, 175
  particular, 54–55, 141
  realized, 133–134, 139–140, 149, 203
  structural content of, 65, 133–134, 140, 166
Regulative ideals, 17, 217, 223
Regulatively correct behavior, 204
Relationship
  any-to-any, 32
  many-to-many, 28, 30–31, 44–45, 55, 220
Replications, 164
Representations, mental, 51, 81–82
Rules, 36–38, 42, 49–52, 54–58, 70, 81, 119, 203–204, 216
  agent-neutral, 52
  general, 85
  public, 51, 64

Science, 1, 11, 38, 45, 47–48, 86–87, 89, 108, 112, 114, 197, 201, 222, 224–225, 229–230
  behavioral, 45, 90, 119, 129, 196, 207, 211
Scientific evidence, 35, 121, 125
Scientific inquiry, 38, 45, 171, 225, 227
Scientific naturalism, 80, 84
Scientific plausibility, 25, 84–85, 135, 144, 188
Scientific reasoning, 32, 43, 86–87, 224
Scientism, 129
Scientists, 11, 13–14, 33, 36–37, 45, 77, 106, 109–110, 121, 134, 170–171, 196, 212, 229
Self, 99, 174, 206–207
  moral, 194
Self-preservation, 29, 48
Self-regard, 205–206
Sentimental order, 209
SIM (social intuitionist model), 87, 108–9, 120–121, 127, 189–190
Situational influences, 211–212
Skepticism, 13, 37, 222
  moral, 70
Skills, 56, 156, 193
Social actions, 38, 52–53, 56, 191, 216

Social behavior, real, 2, 18, 50, 53, 176, 185
Social behaviors, 2–3, 5–7, 10–11, 17–19, 24–25, 43–45, 53, 64–65, 143–144, 155–156, 181–182, 188, 194, 196–197, 226
existing, 9, 17, 155
possible, 49
small-scale, 192
Social goods
small-scale, 193–194
society-wide, 193
Social groups, 4, 164, 194–195
Social institutions, 5, 16, 61, 138–140, 143, 149–154, 158, 160–162, 175–176, 181, 188, 190–191, 205–206, 214, 223–224
existing, 188, 199, 221
local, 139, 157, 161, 172, 188
Social intuitionist model. *See* SIM
Social outcome, 7, 135, 144–145
complex, 135
good, 6, 148, 168, 176, 186, 205
Social psychology, 113, 120, 164, 175, 179, 196, 210–211, 217, 227
Social reality, 172–174
Social research, 159, 168, 182
Social roles, 19, 52, 140
Social sciences, 1–2, 5, 10, 13, 45, 135, 151, 153–159, 160, 174, 176, 219, 221, 228, 231
Social theory, 10, 20, 96, 156, 159–160, 168–169, 176–177, 185–188, 191–192, 195, 205, 223
hypothetical, 187
Social-turn moral psychology, 2, 6–8, 24, 225–226, 228–230
Social worlds, 2, 65, 70, 104, 153–154, 158, 168, 174–176, 184, 203, 223
Society, 19, 33, 36, 45, 59–60, 62, 70–71, 106, 110, 130, 146–147, 157, 173, 178
contemporary, 43, 129
Sociobiological, 127
Sociobiology, 37, 46, 48
Species, 1, 16, 26, 28, 30–31, 33, 45, 136, 144, 170, 178
nonhuman, 45, 143
Sports, 17, 50–53, 56, 76, 81, 216
Stage
approximate-but-useful model, 171
causal descriptive, 170–172, 176
formal model, 171–172
Stories, 11–12, 41–43, 95, 115, 122, 136, 145, 195, 207, 211
Strategies, 9, 14, 37, 49–51, 67, 96, 99, 117, 165, 220
Structural content, 55–62, 64–66, 69–77, 97–98, 104, 140–141, 143, 149, 152, 169, 175–177, 203, 216, 220–222, 225
Structural content of utilitarianism, 57
Structural equation models, 153, 162
Structural explanations, 3, 5, 8, 18–20, 23–25, 37, 44–45, 76, 181
Structural processes, 5–6, 35, 143, 151, 225
Students, 2–4, 17, 82, 165–167, 185, 209
Subjective well-being, 208, 210, 216–217
Success, normative, 203
Systems
cognitive, 30
complex natural, 14, 229
judicial, 2
mind-brain, 96
psychological, 99, 117–118
social, 81
unified natural, 13

Technical concepts, 79, 135, 170
Theoretical perspective, 13–14, 16, 91, 153
Theories
ideal, 61
nonideal, 61–62

Traits, 23–24, 26–28, 30–31, 33, 41–42, 64
  characterological, 205, 211
  psychological, 27
Trolleyology, 92
Trolley problem, 92–93, 95, 99, 112–115
Truth, 30, 50, 65, 80–81, 83, 85–86, 109–110, 119, 125, 135, 173, 182, 215–216, 221, 229
  approximate, 89, 112, 133
  categorical, 172
  literal, 222
  necessary, 11
  possible, 105, 220
  revisable, 224
  universal, 152, 221, 223

Ultimate causes, 31
Ultimate definitions, 41
Ultimate explanations, 8, 23–24, 26–36, 40–42, 44–45, 176
  confirmed, 34
  constructive speculative, 33
  correct, 27
  credible, 44
  scientifically plausible, 28–32, 34, 42, 44
  sophisticated, 41
  speculative, 42
Ultimate functions, 26, 29–30
Ultimate hypothesis, 40–41
Ultimate processes, 32
Ultimate questions, 26, 29
Undergraduates, 79, 91, 97, 121–122, 199
Understanding, 16–17, 27, 32, 36, 50, 59, 77, 89, 131, 151, 192, 199, 212, 225–227
Unhappiness, 161
Unified theory, 69, 217
Unity, homeostatic, 144, 148
Universal external validity, 93–94, 96, 112
Us/them, 190–191
Utilitarianism, 38, 57–58, 67, 74, 76–77, 79, 87, 98–101, 124, 172, 174, 178–179
  economic, 172–174, 176
Utility, 36–37, 46, 71

Valuable outcomes, 6–7, 138, 140, 144–149, 151–152, 154–156, 159, 162, 168–169, 175, 177, 181, 190, 192–193, 199
Valuable social outcomes, 6, 12, 139, 147–148, 159–160, 167–168, 176, 205–206
Value, 7, 19, 26, 126, 135, 146, 173, 193, 205–207, 210
  nonpsychological theory of, 135
  personal, 207–209, 224, 226
Value judgments, 19, 147, 173
Variations, 92–93, 152, 164
Vertices, 186–187
Virtue ethics, 15, 63–64, 67, 80, 217, 226
Virtuous character, 226

Warning calls, 29–30
Wealth, 136, 138–139, 145, 181, 191, 193
WEIRD people, 114, 200
Welfare economics, 172, 176, 178
Well-being, 157–158, 216–217
Wittgenstein, Ludwig, 51, 54, 65
World
  messy, 229
  moral, 111, 215
  possible, 71